HEINEMANN MODULAR MATHEMATICS
for
EDEXCEL AS AND A-LEVEL
Mechanics 4

John Hebborn   J

1

2

3

4

5

Heinemann

Edexcel
Success through qualifications

T

Heinemann Educational Publishers,
a division of Heinemann Publishers (Oxford) Ltd,
Halley Court, Jordan Hill, Oxford, OX2 8EJ

OXFORD   MELBOURNE   AUCKLAND   JOHANNESBUURG
BLANTYRE   GABORONE   PORTSMOUTH NH (USA)   CHICAGO

First published 2001

05   04   03   02

10   9   8   7   6   5   4   3   2

ISBN 0 435 51077 0

Cover design by Gecko Limited

Original design by Geoffrey Wadsley: additional desgn work by Jim Turner

Typeset and illustrated by Tech-Set Ltd, Gateshead, Tyne and Wear

Printed in Great Britain by Scotprint, Haddington

Acknowledgements:

The publisher's and authors' thanks are due to the Edexcel for permission to
reproduce questions from past examination papers. These are marked with an [E].
   The answers have been provided by the authors and are not the responsibility
of the examining board.

About this book

This book is designed to provide you with the best preparation possible for your Edexcel M4 exam. The series authors are senior examiners and exam moderators themselves and have a good understanding of Edexcel's requirements.

Finding your way around

To help to find your way around when you are studying and revising use the:

- **edge marks** (shown on the front page) – these help you to get to the right chapter quickly;
- **contents list** – this lists the headings that identify key syllabus ideas covered in the book so you can turn straight to them;
- **index** – if you need to find a topic the **bold** number shows where to find the main entry on a topic.

Remembering key ideas

We have provided clear explanations of the key ideas and techniques you need throughout the book. Key ideas you need to remember are listed in a **summary of key points** at the end of each chapter and marked like this in the chapters:

■
$$e = \frac{v}{u} \quad \text{or} \quad v = eu$$

Exercises and exam questions

In this book questions are carefully graded so they increase in difficulty and gradually bring you up to exam standard.

- **past exam questions** are marked with an [E];
- **review exercises** on pages 45 and 115 help you practise answering questions from several areas of mathematics at once, as in the real exam;
- **exam style practice paper** – this is designed to help you prepare for the exam itself;
- **answers** are included at the end of the book – use them to check your work.

Contents

5 Stability

Relative motion

When the motions of two particles A and B relative to a fixed point O are known, then the motion of A relative to B or B relative to A can be studied. An example of such a situation which could be modelled by this is when an observer on one ship A, which is moving, observes the motion of a second ship B, which is also moving. The motion that the observer would see is the motion of B relative to A. Another example would be the problem of 'near misses' of aircraft. It is the motion of one aircraft relative to the other that is of primary importance in avoiding a near miss.

Before studying relative motion we will first consider relative positions. Some work on relative position vectors was included in Book M1, Chapter 2.

1.1 Relative position and relative displacement

Suppose particle A has position vector $\mathbf{r}_A$ relative to a fixed point O and particle B has position vector $\mathbf{r}_B$ relative to the same point. Then a vector triangle can be drawn:

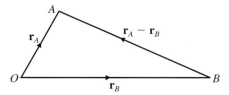

The side BA of the triangle represents the position vector of A relative to B. So

■ **position vector of A relative to $B = \mathbf{r}_A - \mathbf{r}_B$**

The position vector of A relative to B is often denoted by ${}_A\mathbf{r}_B$ giving

$$ {}_A\mathbf{r}_B = \mathbf{r}_A - \mathbf{r}_B $$

If the two particles A and B are moving and at some time $_A\mathbf{r}_B = \mathbf{0}$ then A and B collide. For two particles which do not collide, the magnitude of the relative position vector is a minimum when the particles are closest together.

For two moving particles A and B the modulus of the position vector of A relative to B at time t is the relative displacement of A from B at that time.

■ **relative displacement of A from $B = |\mathbf{r}_A - \mathbf{r}_B|$**

Example 1

James and Kevin are kicking a ball to one another. At time $t = 0$, the ball is at the point with position vector $(6\mathbf{i} + 8\mathbf{j})$ m relative to a fixed point O. James then kicks the ball with constant velocity $(\mathbf{i} - 2\mathbf{j})\,\mathrm{m\,s}^{-1}$. At that moment Kevin is at the point with position vector $2\mathbf{i}$ m and he is running with a constant velocity $(3\mathbf{i} + 2\mathbf{j})\,\mathrm{m\,s}^{-1}$. Show that Kevin will intercept the ball and find the position vector of the point of interception.

The position vector $\mathbf{r}_B$ m of the ball at time t seconds is given by:

$$\mathbf{r}_B = 6\mathbf{i} + 8\mathbf{j} + t(\mathbf{i} - 2\mathbf{j})$$

since it starts from the point with position vector $(6\mathbf{i} + 8\mathbf{j})$ m and moves with velocity $(\mathbf{i} - 2\mathbf{j})\,\mathrm{m\,s}^{-1}$.

Kevin's position vector $\mathbf{r}_K$ m at time t seconds is given by:

$$\mathbf{r}_K = 2\mathbf{i} + t(3\mathbf{i} + 2\mathbf{j})$$

(See the vector equation of a line, Book P3 chapter 5.)

The position vector $_B\mathbf{r}_K$ m of the ball relative to Kevin is therefore given by:

$$_B\mathbf{r}_K = \mathbf{r}_B - \mathbf{r}_K = 6\mathbf{i} + 8\mathbf{j} + t(\mathbf{i} - 2\mathbf{j}) - [2\mathbf{i} + t(3\mathbf{i} + 2\mathbf{j})]$$
$$= (4 - 2t)\mathbf{i} + (8 - 4t)\mathbf{j}$$

Hence $_B\mathbf{r}_K = \mathbf{0}$ when $4 - 2t = 0$ **and** $8 - 4t = 0$, that is when $t = 2$.

So $_B\mathbf{r}_K = \mathbf{0}$ when $t = 2$ and Kevin intercepts the ball at that time.

When $t = 2$,

$$\mathbf{r}_K = 2\mathbf{i} + 2(3\mathbf{i} + 2\mathbf{j})$$
$$= 8\mathbf{i} + 4\mathbf{j}$$

So Kevin intercepts the ball at the point with position vector $(8\mathbf{i} + 4\mathbf{j})$ m.

1.2 Relative velocity

When two particles A and B are moving we have seen that their position vectors are related by

$$_A\mathbf{r}_B = \mathbf{r}_A - \mathbf{r}_B$$

Differentiating this equation with respect to time gives

$$_A\mathbf{v}_B = \mathbf{v}_A - \mathbf{v}_B$$

where $_A\mathbf{v}_B$ is the velocity of A relative to B and $\mathbf{v}_A$ and $\mathbf{v}_B$ are the velocities of A and B respectively.

This relative velocity equation can be re-written to read

$$\mathbf{v}_A =_A \mathbf{v}_B + \mathbf{v}_B$$

or: $$\mathbf{v}_A = \mathbf{v}_B +_A \mathbf{v}_B$$

The velocity triangle is then

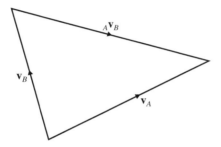

This form of the velocity equation is particularly applicable to problems such as that of a boat being rowed across a river which has a current. The equation shows that the actual velocity of the boat (that is the velocity with which the boat is seen to move) is the resultant of the velocity of the boat relative to the water (that is the speed at which the boat is rowed and the direction in which it is pointed) and the velocity of the water. The actual velocity of an aeroplane experiencing a cross-wind can also be calculated using this equation.

Example 2

A boy who can row at $7\,\text{m s}^{-1}$ in still water is rowing across a river in which the current is flowing at $3\,\text{m s}^{-1}$. Modelling the boat as a particle:

(a) determine at what angle to the river bank the boy must steer his boat if he is to reach the point on the far bank directly opposite his starting position.

Given that the river is $70\,\text{m}$ wide,
(b) calculate the time the boat takes to cross the river.

(a) The speed at which the boy can row in still water gives us the magnitude of the velocity of the boat relative to the water.

Using:
$$_B\mathbf{v}_W + \mathbf{v}_W = \mathbf{v}_B$$

where $\mathbf{v}_B$ is the velocity of the boat and $\mathbf{v}_W$ is the velocity of the water, gives the velocity vector triangle:

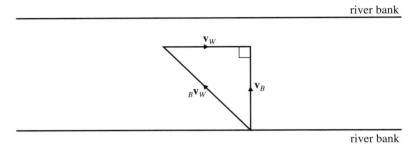

river bank

river bank

This is a right-angled triangle as the boy is moving across the river to the point directly opposite his starting position.

Using the information given, the diagram becomes:

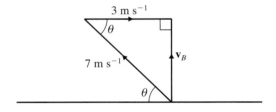

So: $\qquad \cos\theta = \frac{3}{7}$

$$\theta = 64.6° \quad \text{(to 3 significant figures)}$$

The boy must steer the boat an angle of 64.6° to the river bank.

(b) To find the time he takes to cross, we must calculate the magnitude of $\mathbf{v}_B$.

By Pythagoras:

$$|\mathbf{v}_B| = (7^2 - 3^2)\,\text{m s}^{-1} = \sqrt{40}\,\text{m s}^{-1}$$

The time taken to cross the river is given by:

$$\text{time} = \frac{\text{distance}}{\text{speed}}$$

So: $\quad \text{time} = \frac{70}{\sqrt{40}}\,\text{seconds} = 11.1\,\text{seconds} \quad \text{(to 3 significant figures)}$

The boy takes 11.1 seconds to cross the river.

Example 3

Gail walks at a speed of $5\,\text{km}\,\text{h}^{-1}$ due west and Brian runs at a speed of $12\,\text{km}\,\text{h}^{-1}$ in a south-easterly direction. Modelling Brian and Gail as particles:

(a) find the velocity of Brian relative to Gail,
(b) find the speed of Brian relative to Gail and the direction of the relative velocity.

The information given can be summarised by the following diagrams:

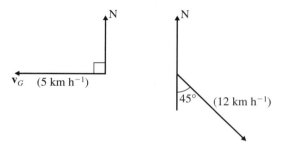

Suppose the unit vector due South is **i** and the unit vector due East is **j**.

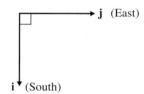

Then the unit vector in the South-East direction is $\frac{1}{\sqrt{2}}(\mathbf{i}+\mathbf{j})$.

So: $\qquad \mathbf{v}_G = -5\mathbf{j}\,\text{km}\,\text{h}^{-1}$

And: $\qquad \mathbf{v}_B = 12 \times \frac{1}{\sqrt{2}}(\mathbf{i}+\mathbf{j})\,\text{km}\,\text{h}^{-1} = 6\sqrt{2}(\mathbf{i}+\mathbf{j})\,\text{km}\,\text{h}^{-1}.$

(a) The velocity of Brian relative to Gail is:

$$_B\mathbf{v}_G = \mathbf{v}_B - \mathbf{v}_G = [6\sqrt{2}(\mathbf{i}+\mathbf{j}) + 5\mathbf{j}]\,\text{km}\,\text{h}^{-1}$$

$$= [6\sqrt{2}\mathbf{i} + (6\sqrt{2}+5)\mathbf{j}]\,\text{km}\,\text{h}^{-1}.$$

(b) The speed of Brian relative to Gail is the modulus of this velocity. So:

$$|_B\mathbf{v}_G| = [(6\sqrt{2})^2 + (6\sqrt{2}+5)^2]^{\frac{1}{2}}\,\text{km}\,\text{h}^{-1}$$

$$= [72 + 72 + 25 + 60\sqrt{2}]^{\frac{1}{2}}\,\text{km}\,\text{h}^{-1}$$

$$= [169 + 60\sqrt{2}]^{\frac{1}{2}}\,\text{km}\,\text{h}^{-1}$$

$$= 15.9\,\text{km}\,\text{h}^{-1}$$

Let the relative velocity make an angle θ with **i**.

Then:

And:
$$\tan \theta = \frac{6\sqrt{2} + 5}{6\sqrt{2}}$$
$$\theta = 57.8°$$

The speed of Brian relative to Gail is $15.9 \, \text{km h}^{-1}$ and the relative velocity is in the direction $S \, 57.8° \, E$.

Example 4

To Fiona, cycling due north at $40 \, \text{km h}^{-1}$, the wind appears to be blowing from the East. To Guy, cycling due South at $50 \, \text{km h}^{-1}$, the wind appears to be blowing from the South-East. Find the actual velocity of the wind.

The given information about the velocities can be summarised in tables:

For Fiona (F) we have:

Velocity	Speed (km h^{-1})	Direction
$\mathbf{v}_F$	40	North
$_w\mathbf{v}_F$		West (from the East)

Using $\mathbf{v}_w = \mathbf{v}_F + {}_w\mathbf{v}_F$ gives the velocity vector triangle:

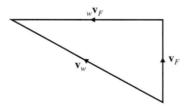

For Guy (G) we have:

Velocity	Speed (km h^{-1})	Direction
$\mathbf{v}_G$	50	South
$_w\mathbf{v}_G$		North-West (from the South-East)

The velocity vector triangle is:

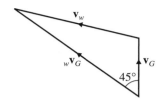

The actual velocity of the wind, $\mathbf{v}_w$, must be the same in both diagrams. In order to emphasise this fact the two diagrams can be combined:

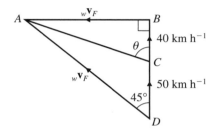

Using $\triangle ABD$: $AB = 90 \tan 45° = 90$

So: $|_w\mathbf{v}_F| = 90 \, \text{km h}^{-1}$

Using $\triangle ABC$: $AC = \sqrt{(40^2 + 90^2)} = 98.5$

And: $\tan \theta = \frac{90}{40} = \frac{9}{4}$

$\theta = 66.0°$ (to 3 s.f.)

So the velocity of the wind is $98.5 \, \text{km h}^{-1}$ in a direction $(360 - 66.0)° = 294°$.

Exercise 1A

1 Relative to a fixed origin O, two particles A and B have position vectors $\mathbf{r}_A$ and $\mathbf{r}_B$ respectively. Find the position vector of A relative to B in the following cases:
   (a) $\mathbf{r}_A = (2\mathbf{i} - 5\mathbf{j}) \, \text{m}$, $\mathbf{r}_B = (6\mathbf{i} + 4\mathbf{j}) \, \text{m}$
   (b) $\mathbf{r}_A = 6\mathbf{i} \, \text{m}$, $\mathbf{r}_B = (3\mathbf{i} + 7\mathbf{j}) \, \text{m}$
   (c) $\mathbf{r}_A = (2\mathbf{i} + 4\mathbf{j} - 3\mathbf{k}) \, \text{m}$, $\mathbf{r}_B = (7\mathbf{i} + 4\mathbf{j} - 2\mathbf{k}) \, \text{m}$

2 $\mathbf{v}_A$ and $\mathbf{v}_B$ are the velocity vectors of two particles A and B. Find the velocity vector of A relative to B and the speed of A relative to B in the following cases:
   (a) $\mathbf{v}_A = (7\mathbf{i} + 2\mathbf{j}) \, \text{m s}^{-1}$, $\mathbf{v}_B = (2\mathbf{i} - 10\mathbf{j}) \, \text{m s}^{-1}$
   (b) $\mathbf{v}_A = 8\mathbf{i} \, \text{m s}^{-1}$, $\mathbf{v}_B = (4\mathbf{i} - 3\mathbf{j}) \, \text{m s}^{-1}$
   (c) $\mathbf{v}_A = (4\mathbf{i} - 2\mathbf{j} + 6\mathbf{k}) \, \text{m s}^{-1}$, $\mathbf{v}_B = (7\mathbf{i} + 5\mathbf{j} - 3\mathbf{k}) \, \text{m s}^{-1}$

3 A and B are two aircraft. A has a velocity of $(400\mathbf{i} + 120\mathbf{j})\,\mathrm{m\,s}^{-1}$.
The pilot of B sees A to have a velocity of $(200\mathbf{i} - 350\mathbf{j})\,\mathrm{m\,s}^{-1}$.
Find the velocity of B in vector form.

4 L and M are two liners. At 1500 h the position vector of the
bow of L relative to a fixed origin O is $(3\mathbf{i} + 5\mathbf{j})\,\mathrm{km}$ and the
position vector of the stern of M relative to O is $(5\mathbf{i} - 4\mathbf{j})\,\mathrm{km}$.
The velocities of L and M are $(6\mathbf{i} - \mathbf{j})\,\mathrm{km\,h}^{-1}$ and
$(4\mathbf{i} + 8\mathbf{j})\,\mathrm{km\,h}^{-1}$ respectively. Show that if both liners maintain
their velocities a collision will occur. Find the time when they
collide and the position vector of the point where they collide.

5   A girl can row a boat at $3\,\mathrm{m\,s}^{-1}$ in still water. She wishes to
cross a river in which there is a current of $1.5\,\mathrm{m\,s}^{-1}$ by the
shortest route possible. The river is 165 m wide. Find
(a) the direction in which she must steer the boat, and
(b) the time she takes to row across the river.

6   A boy can row his boat at $7\,\mathrm{m\,s}^{-1}$ in still water. He rows with
his boat pointing in a direction at an angle of $40°$ to a river
bank in the upstream direction. Given that the river is flowing
at $5\,\mathrm{m\,s}^{-1}$ find the speed and direction in which the boy and
his boat actually travel.

7   A bird is capable of flying at $80\,\mathrm{km\,h}^{-1}$. It wishes to fly to its
nest which is due East of its present position. There is a wind
blowing from the North-West at $70\,\mathrm{km\,h}^{-1}$. Find the direction
in which the bird must fly to reach its nest.

8   A passenger on a train which is travelling due North at $75\,\mathrm{km\,h}^{-1}$
observes a car. The car appears to be moving at $60\,\mathrm{km\,h}^{-1}$ in the
direction N 60° E. Find the true velocity of the car.

9   To a man travelling due North at $16\,\mathrm{km\,h}^{-1}$ the wind
appeared to blow from the West. When he doubled his speed
to $32\,\mathrm{km\,h}^{-1}$ the wind appeared to blow from the North-West.
Find the velocity of the wind.

10   A boat is sailing West at a speed of $12\,\mathrm{km\,h}^{-1}$. To a man on
the boat the wind appeared to blow from a direction $160°$.
A second boat is sailing East at a speed of $15\,\mathrm{km\,h}^{-1}$. To a
woman on that boat the wind appeared to be blowing from a
direction $120°$. Find the velocity of the wind.

1.3 Interception and collision

For two moving people, A and B, the velocity of A relative to B tells us how A moves relative to B, in other words, how B sees A move. If A and B meet at a point, B will see A moving directly towards him. (A will also see B moving directly towards her.)

Suppose A_0 is the initial position of A, with position vector $\mathbf{r}_{A_0}$, and B_0 is the initial position of B, with position vector $\mathbf{r}_{B_0}$. Then at time t the position vectors of A and B are:

$$A: \qquad \mathbf{r}_A = \mathbf{r}_{A_0} + \mathbf{v}_A t,$$

$$B: \qquad \mathbf{r}_B = \mathbf{r}_{B_0} + \mathbf{v}_B t.$$

If A and B collide then there exists a value of t for which $\mathbf{r}_A = \mathbf{r}_B$ that is:

$$\mathbf{r}_{A_0} + \mathbf{v}_A t = \mathbf{r}_{B_0} + \mathbf{v}_B t$$

Rearranging gives:

$$\mathbf{r}_{B_0} - \mathbf{r}_{A_0} = t(\mathbf{v}_A - \mathbf{v}_B) = t\,_A\mathbf{v}_B.$$

This means that the relative velocity $_A\mathbf{v}_B$ is parallel to the line $\overrightarrow{A_0 B_0}$ which is the line joining A_0, the initial position of A, to B_0, the initial position of B.

Hence, for interception or collision the velocity of A relative to B is directed from the initial position A_0 of A to the initial position B_0 of B. That is:

- **For interception, $_A\mathbf{v}_B$ is directed along the line joining the initial position A_0 of A to the initial position B_0 of B.**

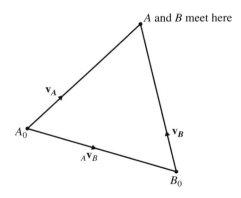

This fact, together with the relative velocity equation

$$\mathbf{v}_A = \mathbf{v}_B + {}_A\mathbf{v}_B$$

enables the course required for interception or collision to be determined.

Example 5

A yacht is sailing at a constant velocity of 3 knots on a bearing 125°. A man in a motor boat sees the yacht when it is 1 nautical mile due North of him. Assuming the motor boat travels at its maximum speed of 5 knots, find:

(a) the direction in which the motor boat must travel in order to intercept the yacht,
(b) the time taken by the motor boat to intercept the yacht.
(Note: 1 knot is 1 nautical mile per hour.)

In order for the motor boat to intercept the yacht the velocity of the boat relative to the yacht must be directed along the line initially joining the boat to the yacht. Originally the yacht is due North of the motor boat. Therefore $_{MB}\mathbf{v}_Y$ must be directed due North.

The information given about the velocities can be summarised in a table:

Velocity	Speed	Bearing
$\mathbf{v}_Y$	3 knots	125°
$\mathbf{v}_{MB}$	5 knots	
$_{MB}\mathbf{v}_Y$		0° (North)

There are two missing quantities in the table. The direction of $\mathbf{v}_{MB}$ is required for part (a) and the magnitude of $_{MB}\mathbf{v}_Y$ is required for part (b).

Using $\mathbf{v}_{MB} = \mathbf{v}_Y + {}_{MB}\mathbf{v}_Y$ gives the velocity vector triangle:

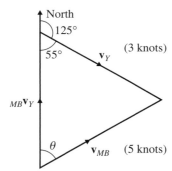

where θ is the angle required.

(a) Using the sine rule gives:

$$\frac{5}{\sin(180° - 125°)} = \frac{3}{\sin\theta}$$

$$\sin\theta = \frac{3\sin 55°}{5}$$

$$\theta = 29.4°$$

The motor boat must travel in a direction 029.4°

(b) The third angle of the velocity triangle is

$$180° - (55° + 29.4°) = 95.6°.$$

Using the sine rule once more gives:

$$\frac{5}{\sin 55°} = \frac{|_{MB}\mathbf{v}_Y|}{\sin 95.6°}$$

$$|_{MB}\mathbf{v}_Y| = \frac{5 \sin 95.6°}{\sin 55°}$$

$$|_{MB}\mathbf{v}_Y| = 6.074 \text{ knots}$$

But the yacht was initially 1 nautical mile from the motor boat.

Hence the time to interception is $\dfrac{1}{6.074} = 0.1646$ h or 9.88 min.

1.4 Closest approach

For any two moving particles which do not meet at a point there will be some instant when they are closest together. This will be when the magnitude of the position vector of A relative to B is a minimum.

That is, when $|_A\mathbf{r}_B| = |_B\mathbf{r}_A| = |\mathbf{r}_B - \mathbf{r}_A|$ is minimum.

The next example uses this fact to find the time when the two particles are closest together and the distance between them at this time.

Example 6

A and B are two small yachts. At 1400 hours B has position vector $(3\mathbf{i} + 6\mathbf{j})$ km relative to A. Yacht A has a constant velocity of $(4\mathbf{i} + 6\mathbf{j})$ km h^{-1} and yacht B has a constant velocity of $(5\mathbf{i} - 2\mathbf{j})$ km h^{-1}. By modelling the yachts as particles find, to the nearest minute, the time when A and B are closest together and the distance between them at this time.

Let 1400 hours be time $t = 0$.

The velocity of B relative to A, $_B\mathbf{v}_A$ km h^{-1}, is given by:

$$\begin{aligned}
_B\mathbf{v}_A &= \mathbf{v}_B - \mathbf{v}_A \\
&= (5\mathbf{i} - 2\mathbf{j}) - (4\mathbf{i} + 6\mathbf{j}) \\
&= \mathbf{i} - 8\mathbf{j}
\end{aligned}$$

Then, at time t hours the position vector of B relative to A, $_B\mathbf{r}_A$ km, is given by:

$$\begin{aligned}
_B\mathbf{r}_A &= (3\mathbf{i} + 6\mathbf{j}) + t(\mathbf{i} - 8\mathbf{j}) \\
&= (3 + t)\mathbf{i} + (6 - 8t)\mathbf{j}
\end{aligned}$$

So:
$$|_B\mathbf{r}_A| = \sqrt{[(3+t)^2 + (6-8t)^2]}$$

$|_B\mathbf{r}_A|$ is minimum when $(3+t)^2 + (6-8t)^2$ is minimum.

Let:
$$\begin{aligned}
f(t) &= (3+t)^2 + (6-8t)^2 \\
&= 9 + 6t + t^2 + 36 - 96t + 64t^2 \\
&= 45 - 90t + 65t^2
\end{aligned}$$

Differentiating with respect to t gives:

$$\frac{\mathrm{d}}{\mathrm{d}t}f(t) = -90 + 130t$$

For closest approach, $\dfrac{\mathrm{d}}{\mathrm{d}t}f(t) = 0$.

But $\dfrac{\mathrm{d}}{\mathrm{d}t}f(t) = 0$ when $t = \frac{90}{130} = \frac{9}{13}$ hours = 41.5 minutes

As $t = 0$ at 1400 hours, the yachts are closest together at 1442 hours (to the nearest minute).

When $t = \frac{9}{13}$,

$$\begin{aligned}
|_B\mathbf{r}_A| &= \sqrt{\left[\left(3 + \tfrac{9}{13}\right)^2 + \left(6 - \tfrac{72}{13}\right)^2\right]} \\
&= 3.72
\end{aligned}$$

The distance between the yachts at 1442 hours is 3.72 km.

Alternatively, the time when the minimum value of $f(t)$ occurs can be found by completing the square:

From above:
$$\begin{aligned}
f(t) &= 45 - 90t + 65t^2 \\
&= 65\left(t^2 - \tfrac{90}{65}t\right) + 45 \\
&= 65\left\{\left(t - \tfrac{45}{65}\right)^2 - \left(\tfrac{45}{65}\right)^2\right\} + 45 \\
&= 65\left(t - \tfrac{45}{65}\right)^2 - 65\left(\tfrac{45}{65}\right)^2 + 45
\end{aligned}$$

The minimum value of $f(t)$ occurs when $t = \frac{45}{65} = \frac{9}{13}$ h = 41.5 min.

As above, this shows that the yachts are closest together at 1442 hours (to the nearest minute).

Substituting $t = \frac{9}{13}$ gives:
$$\begin{aligned}
f\left(\tfrac{9}{13}\right) &= -65\left(\tfrac{9}{13}\right)^2 + 45 \\
&= 45 - \frac{5 \times 81}{13}
\end{aligned}$$

and as:
$$f(t) = |_B\mathbf{r}_A|^2$$

this gives:
$$|_B\mathbf{r}_A|_{\min} = \sqrt{\left(45 - \frac{5 \times 81}{13}\right)} = 3.72$$

So the minimum distance between the yachts is 3.72 km.

Closest approach using a scalar product

As we have seen, two moving particles A and B are closest together when:

$$|_A\mathbf{r}_B| \text{ is a minimum.}$$

But $|_A\mathbf{r}_B|$ is a minimum when $|_A\mathbf{r}_B|^2$ is a minimum and $|_A\mathbf{r}_B|^2 = {_A\mathbf{r}_B} \cdot {_A\mathbf{r}_B}$ (Book P3 chapter 5).

Differentiating this equation with respect to time gives

$$\frac{\mathrm{d}}{\mathrm{d}t}\left(|_A\mathbf{r}_B|^2\right) = 2{_A\mathbf{r}_B} \cdot {_A\mathbf{v}_B}$$

(by the product rule, and since $|_A\mathbf{r}_B|$ is minimum when $\frac{\mathrm{d}}{\mathrm{d}t}\left(|_A\mathbf{r}_B|^2\right) = 0$ this shows that $|_A\mathbf{r}_B|$ is minimum when:

$${_A\mathbf{r}_B} \cdot {_A\mathbf{v}_B} = 0$$

This gives us an alternative method of solving closest approach problems which is often more efficient, particularly when $_A\mathbf{r}_B$ and $_A\mathbf{v}_B$ are given in vector form.

■ $|_A\mathbf{r}_B|$ **is a minimum when** $_A\mathbf{r}_B \cdot {_A\mathbf{v}_B} = 0$

Example 7

Use the scalar product method to find the time when the two yachts of example 6 are closest together.

As before:
$$_B\mathbf{v}_A = \mathbf{i} - 8\mathbf{j}$$

and:
$$_B\mathbf{r}_A = (3 + t)\mathbf{i} + (6 - 8t)\mathbf{j}$$

So:
$$_B\mathbf{r}_A \cdot {_B\mathbf{v}_A} = [(3 + t)\mathbf{i} + (6 - 8t)\mathbf{j}].(\mathbf{i} - 8\mathbf{j})$$

$$= (3 + t) - 8(6 - 8t)$$

$$= -45 + 65t$$

$_B\mathbf{r}_A \cdot {_B\mathbf{v}_A} = 0$ when $t = \frac{45}{65} = \frac{9}{13}$ h $= 41.5$ minutes

Therefore they are closest together at 1442 hours.

Questions are not always set in vector form. Such an example is considered on the following page.

Example 8

At noon a ship P is $50\,\text{km}$ West of a ship Q. P is travelling at $12\,\text{km h}^{-1}$ in a direction $050°$ and Q is travelling at $16\,\text{km h}^{-1}$ in a direction $330°$. Find the least distance between the two ships in the subsequent motion and the time when they are closest together.

The given information about the velocities is summarised in the table:

Velocity	Magnitude	Direction
$\mathbf{v}_P$	$12\,\text{km h}^{-1}$	$050°$
$\mathbf{v}_Q$	$16\,\text{km h}^{-1}$	$320°$

Use $\mathbf{v}_P = \mathbf{v}_Q + {}_P\mathbf{v}_Q$ to obtain the velocity vector triangle (as in example 4).

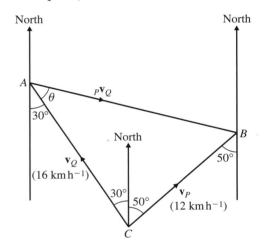

Using the cosine rule in $\triangle ABC$ gives:

$$|{}_P\mathbf{v}_Q|^2 = 16^2 + 12^2 - 2 \times 16 \times 12 \cos 80°$$

$$|{}_P\mathbf{v}_Q| = 18.26 \tag{1}$$

Using the sine rule in $\triangle ABC$ gives:

$$\frac{12}{\sin \theta} = \frac{18.26}{\sin 80°}$$

$$\sin \theta = \frac{12 \sin 80°}{18.26}$$

$$\theta = 40.34° \tag{2}$$

Hence, using (2), the direction of ${}_P\mathbf{v}_Q$ is

$$180° - (30° + 40.34°) = 109.66° \tag{3}$$

Now consider the motion of P relative to Q. The relative path is along the direction of $_P\mathbf{v}_Q$. Initially P is 50 km west of Q so the displacement diagram is:

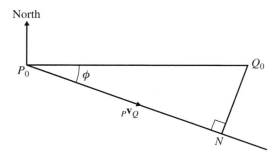

The shortest distance between P and Q is given by the perpendicular distance from Q_0 onto the direction of travel of P relative to Q, that is $Q_0 N$.

Using (3) gives: $\phi = 109.66° - 90° = 19.66°$

and $Q_0 N = 50 \sin \phi$

$$= 50 \sin 19.66°$$

$$= 16.82$$

So the least distance between the ships is 16.8 km.

To find the time when this occurs the distance $P_0 N$ must be found.

$$P_0 N = 50 \cos \phi$$

$$= 50 \cos 19.66°$$

$$= 47.09$$

Therefore, using (1), the time to reach the point of closest approach is

$$\frac{P_0 N}{|_P\mathbf{v}_Q|} = \frac{47.09}{18.26} = 2.578 \, \text{h} = 2 \, \text{h} \, 35 \, \text{min.}$$

So P and Q are closest together at 1435 hours.

Course needed for closest approach

The above example shows how to find the shortest distance between two moving particles A and B when their velocities are given. It is also possible to find the necessary direction of travel for one particle A to pass as close as possible to a second particle B.

Suppose that A and B are two particles initially at A_0 and B_0. The velocity of A is $\mathbf{v}_A$, a known constant vector. However, only the constant speed of B is known, not its direction of travel. We wish to find the direction in which B must travel so that B will pass as close to A as possible. The problem is best approached by drawing a combined displacement and velocity diagram:

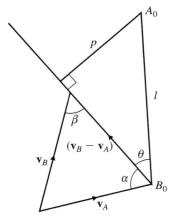

The velocity vector triangle is:

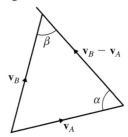

and the displacement vector triangle is:

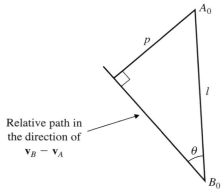

The displacement vector triangle shows that B will pass as close as possible to A when p is as **small** as possible. Since $p = l \sin \theta$ this means that θ is as **small** as possible.

For all directions of travel of B the following remain unchanged:

(i)   the speed of A, the speed of B, the length of $A_0 B_0 = l$,

(ii) the direction of the velocity of A and the direction of $\overrightarrow{A_0 B_0}$.

As a consequence of (ii) the angle between the speed of A and the direction of $\overrightarrow{A_0 B_0}$, which is $(\alpha + \theta)$, remains unchanged.

Since $(\alpha + \theta)$ is constant and θ is as **small** as possible it follows that α is as **large** as possible. Using the sine rule in the velocity vector triangle gives:

$$\frac{|\mathbf{v}_A|}{\sin \beta} = \frac{|\mathbf{v}_B|}{\sin \alpha}$$

So:

$$\sin \alpha = \frac{|\mathbf{v}_B|}{|\mathbf{v}_A|} \sin \beta$$

As the direction of travel of B varies, $\sin \alpha$ is largest when $\sin \beta = 1$, that is when $\beta = 90°$.

■ **For B to pass as close as possible to A the direction of motion of B should be perpendicular to the relative path, that is to $(\mathbf{v}_B - \mathbf{v}_A)$.**

Example 9 illustrates the use of this fact when solving a problem.

Example 9

A and B are two yachts. A is sailing at a constant speed of $15\,\text{km h}^{-1}$ on a bearing of $030°$. At noon the yachtswoman on B sees A $4\,\text{km}$ to the West. B is sailing at a constant speed of $12\,\text{km h}^{-1}$.

(a) Show that it is not possible for B to intercept A.
(b) Find the course that B must set in order to pass as close as possible to A.

(a) Assume that B can intercept A. Then $_B\mathbf{v}_A$ will be directed due West as this is the direction of the line joining the initial positions of A and B.

The known information about the velocities is:

Velocity	Speed	Bearing
$\mathbf{v}_A$	$15\,\text{km h}^{-1}$	$030°$
$\mathbf{v}_B$	$12\,\text{km h}^{-1}$	
$_B\mathbf{v}_A$		$270°$ (West)

Use the available information and $\mathbf{v}_B = \mathbf{v}_A + {}_B\mathbf{v}_A$ to draw the velocity triangle (as in example 4).

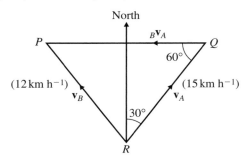

The perpendicular from R onto PQ has length

$$15 \cos 30° = 15 \frac{\sqrt{3}}{2} = 12.99$$

As this is greater than the length of PR the triangle cannot be drawn. Hence B cannot intercept A.

(b) The known information about the velocities is:

Velocity	Speed	Bearing
$\mathbf{v}_A$	$15\,\text{km h}^{-1}$	$030°$
$\mathbf{v}_B$	$12\,\text{km h}^{-1}$	

For B to pass as close as possible to A, $\mathbf{v}_B$ must be perpendicular to $_B\mathbf{v}_A$.

Also: $$\mathbf{v}_B = \mathbf{v}_A + {}_B\mathbf{v}_A$$

So the velocity vector triangle is:

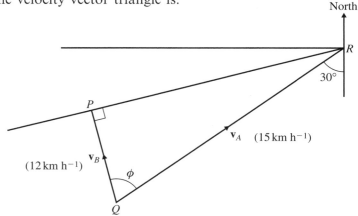

(Note that the triangle is unique as the hypotenuse and one other side are known.)

Hence: $$\cos \phi = \tfrac{12}{15}, \; \phi = 36.9°$$

So the direction of $\mathbf{v}_B$ is $360° - (36.9° - 30°) = 353°$ (3 s.f.).

Exercise 1B

1 P and Q are two aircraft. At 12 noon, the position vector of the pilot of P relative to a fixed origin O is $(5\mathbf{i} + 7\mathbf{j})\,\text{km}$ and that of the pilot of Q relative to O is $(115\mathbf{i} - 25\mathbf{j})\,\text{km}$. The constant velocity vectors of P and Q are $(100\mathbf{i} - 285\mathbf{j})\,\text{km h}^{-1}$ and $(75\mathbf{i} - 135\mathbf{j})\,\text{km h}^{-1}$ respectively. Calculate the closest distance between the two pilots and the time to the nearest minute when they are this distance apart.

2   Peter is running with speed $7.5 \, \text{m s}^{-1}$ in the direction $-3\mathbf{i} + 4\mathbf{j}$ and Jayesh is running with speed $6.5 \, \text{m s}^{-1}$ in the direction $12\mathbf{i} - 5\mathbf{j}$. At time $t = 0$ Peter passes through the point with position vector $(17\mathbf{i} + 5\mathbf{j}) \, \text{m}$ and Jayesh passes through the point with position vector $(4\mathbf{i} + 20\mathbf{j}) \, \text{m}$ both relative to a fixed origin O.
   (a) Find the time in seconds when Peter and Jayesh are closest together and the distance they are apart at this time.
   (b) When $t = 1 \, \text{s}$, Jayesh alters his velocity so that he intercepts Peter when $t = 2 \, \text{s}$. Find his new velocity in vector form.

3   At time $t = 0$ A and B pass through the points with position vectors $(3\mathbf{i} + 4\mathbf{j} - 2\mathbf{k}) \, \text{m}$ and $(2\mathbf{i} - 5\mathbf{j} - \mathbf{k}) \, \text{m}$ relative to a fixed origin O. A has velocity $(2\mathbf{i} + 2\mathbf{j} - 5\mathbf{k}) \, \text{m s}^{-1}$ and B has velocity $(2.5\mathbf{i} + p\mathbf{j} + q\mathbf{k}) \, \text{m s}^{-1}$ where p and q are constants. A and B subsequently collide. Find:
   (a) the time when they collide,
   (b) the values of p and q,
   (c) the position vector of the point of collision.

4   Two cars are travelling along two straight roads which cross at right angles. At 10 a.m. one car is approaching the junction at $20 \, \text{m s}^{-1}$ and is 200 m away from the junction. At the same time, the second car is travelling along the other road towards the junction at a speed of $25 \, \text{m s}^{-1}$ and is 100 m from the junction. Assuming that neither car changes its speed or direction during the subsequent motion, calculate:
   (a) the magnitude of the velocity of one car relative to the other,
   (b) the shortest distance between the cars in the subsequent motion.

5   P and Q are two motor boats. P is travelling at a constant speed of $20 \, \text{km h}^{-1}$ on a bearing of $060°$ and Q is travelling at a constant speed of $18 \, \text{km h}^{-1}$. At 2 p.m. a man in Q sees P 4 km away to the West. Find:
   (a) the course that Q must set in order to pass as close as possible to P,
   (b) the shortest distance between P and Q in the subsequent motion,
   (c) the time, to the nearest minute, when the distance between the two boats is least.

6 Jodie is walking at a constant speed of $6 \, \text{km h}^{-1}$ on a bearing of 030°. Rowena is walking at a constant speed of $5 \, \text{km h}^{-1}$ on a bearing of 280°. At noon Rowena is 3 km East of Jodie. Assuming neither girl changes her velocity find:
(a) the velocity of Jodie relative to Rowena,
(b) the shortest distance between the two girls in the subsequent motion.

7 Car A is travelling along a straight road on a bearing of 300° at a constant speed of $80 \, \text{km h}^{-1}$. At 10 a.m. it is 15 km due East of a second car B, which is also travelling at a constant speed. Assuming that car B can travel in any direction,
(a) find the minimum speed of B if it is to intercept A.
Given that car B is travelling at a speed of $50 \, \text{km h}^{-1}$,
(b) find, to the nearest minute, the earliest time at which it can intercept A.

8 At 10 a.m. a battleship's radar sights an enemy cruiser to the North. The battleship has a maximum speed of $35 \, \text{km h}^{-1}$ and the cruiser is travelling at a constant velocity of $45 \, \text{km h}^{-1}$ on a bearing of 120°. Assuming the battleship travels at its maximum speed in pursuit of the cruiser,
(a) find the course that the battleship must set in order to pass as close as possible to the cruiser.
Given that the cruiser was initially sighted 60 km from the battleship, and the battleship's guns have a maximum range of 20 km,
(b) determine whether the battleship will be able to get within range of the cruiser.

9 At noon ship A is 10 km West of a second ship B. Ship A is travelling due South at a constant speed of $15 \, \text{km h}^{-1}$. Ship B is moving at a constant speed of $12 \, \text{km h}^{-1}$. Find:
(a) the course B should set in order to pass as close as possible to A,
(b) the shortest distance between the two ships in the subsequent motion,
(c) the time when they are closest together, giving your answer to the nearest minute.

SUMMARY OF KEY POINTS

1 The position vector of A relative to B, $_A\mathbf{r}_B$, is given by:

$$_A\mathbf{r}_B = \mathbf{r}_A - \mathbf{r}_B$$

where $\mathbf{r}_A$ and $\mathbf{r}_B$ are the position vectors of A and B relative to a fixed origin.

2 Two moving particles A and B collide if at some time $_A\mathbf{r}_B = 0$.

3 The velocity of A relative to B, $_A\mathbf{v}_B$, is given by:

$$_A\mathbf{v}_B = \mathbf{v}_A - \mathbf{v}_B$$

where $\mathbf{v}_A$ and $\mathbf{v}_B$ are the actual velocities of A and B.

4 For two particles A and B to intercept or collide $_A\mathbf{v}_B$ is directed along the line joining their initial positions.

5 Two particles A and B are closest together when $|_A\mathbf{r}_B|$ has a minimum value. This minimum can be found by differentiating $|_A\mathbf{r}_B|^2$ with respect to time.

6 Alternatively, $|_A\mathbf{r}_B|$ is a minimum when $_A\mathbf{r}_B \cdot {}_A\mathbf{v}_B = 0$.

7 In order for B to pass as close as possible to A the direction of motion of B should be perpendicular to the relative path.

Elastic collisions in two dimensions

Chapter 4 of Book M2 was concerned with 'collisions'. However, the treatment there was restricted to one-dimensional problems such as the direct impact of two particles moving in the same straight line and normal impact of a particle with a fixed surface. Such an impact is unusual in real situations. So in this chapter we shall consider the generalisation of that work to elastic collisions in two-dimensional situations.

Assume that the elastic bodies involved in the collisions are **smooth**. This means that the mutual reaction acts along **the common normal at the point of impact**. For example, when two elastic spheres collide the mutual reaction acts along **the line of centres**.

By the impulse–momentum principle there is **no change in the momentum** of the bodies involved in the collision **perpendicular** to this common normal and hence:

■ **the components of the velocities of the bodies involved in the collision perpendicular to the common normal are unchanged.**

The components of the velocities of the bodies involved in the collision **parallel to this common normal** can be found in the same manner as for direct impact. That is:

(i) total momentum in this direction before impact
= total momentum in this direction after impact

(ii) $\dfrac{\text{speed of separation in this direction}}{\text{speed of approach in this direction}} = e$

where e is the coefficient of restitution between the bodies (Newton's law of restitution).

So it follows that, if you have an elastic collision, you must deal with the motions along the common normal and perpendicular to it separately. In the particular case of two elastic spheres colliding, this means that the motions along and perpendicular to the line of centres must be dealt with separately. You will see this clearly in the next two sections when we consider the cases of oblique impact of a smooth sphere with a fixed smooth surface and oblique impact of two smooth spheres.

2.1 Impact of a smooth sphere with a fixed smooth plane

Consider a smooth sphere that strikes a fixed smooth plane *obliquely*, that is, not along the normal to the plane. Suppose that the sphere has speed u and that its direction of motion makes an angle α with the fixed plane. Since the sphere is smooth and the plane is smooth the mutual reaction between the sphere and the plane will act along the normal to the plane at the point of impact. So we can consider the motions parallel to the plane and perpendicular to the plane separately. Just before impact we have:

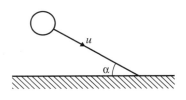

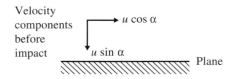

Suppose that after the impact the sphere has speed v and that its direction of motion makes an angle β with the fixed plane:

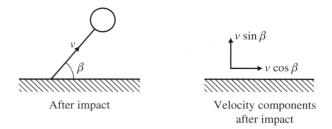

After impact   Velocity components after impact

(i) Perpendicular to the common normal, that is, parallel to the plane, there is no change in the component of the velocity so that

$$u \cos \alpha = v \cos \beta \qquad (1)$$

(ii) Perpendicular to the plane:

$$\frac{\text{component of velocity in this direction after collision}}{\text{component of velocity in this direction before collision}} = e$$

or

(component of velocity after collision) $= e$(component of velocity before collision)

$$v \sin \beta = eu \sin \alpha \qquad (2)$$

You can use the solutions of equations (1) and (2) to calculate the changes in the speed and direction of motion of the sphere and also the change in momentum and loss of kinetic energy as a result of the impact.

Example 1

A small smooth sphere of mass 2 kg is projected along a smooth horizontal table towards a fixed smooth vertical wall. Before impact with the wall it has a speed of $12\,\text{m s}^{-1}$ and its direction of motion makes an angle of $30°$ with the wall. Given that the coefficient of restitution between the sphere and the wall is $\frac{1}{4}$,

(a) calculate the velocity of the sphere after the impact.
(b) Find also the magnitude of the impulse exerted by the wall on the sphere.

(a) The situation *before impact* is:

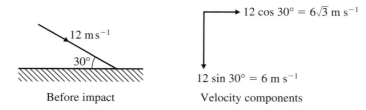

| Before impact | Velocity components |

Suppose that the final speed is $v\,\text{m s}^{-1}$ and the direction of motion makes an angle β with the wall. The situation *after impact* is:

| After impact | Velocity components |

(i) *Parallel to the wall*
No change in velocity component, so: $6\sqrt{3} = v\cos\beta$ (1)

(ii) *Perpendicular to the wall*
Newton's law of restitution gives: $v\sin\beta = \frac{1}{4} \times 6 = 1\frac{1}{2}$ (2)

Squaring and adding equations (1) and (2) gives:

$$v^2\sin^2\beta + v^2\cos^2\beta = v^2 = (6\sqrt{3})^2 + (1\tfrac{1}{2})^2$$
$$= 110\tfrac{1}{4}$$

So: $$v = 10\tfrac{1}{2}$$

Dividing (2) by (1) gives:

$$\frac{v\sin\beta}{v\cos\beta} = \tan\beta = \frac{1\frac{1}{2}}{6\sqrt{3}} = \frac{\sqrt{3}}{12}$$

So: $$\beta = 8.21°$$

So the sphere has speed $10\frac{1}{2}\,\text{m s}^{-1}$ and the direction of motion makes an angle of $8.21°$ with the wall.

(b) Now consider the motion of the sphere perpendicular to the wall.

Impulse exerted by the wall on the sphere

$$= \text{change in momentum of the sphere}$$

$$= 2v \sin \beta + 2(12 \sin 30°)$$

$$= 2(1\tfrac{1}{2}) + 2(6) = 15$$

So the magnitude of the impulse exerted by the wall on the sphere is 15 Ns.

Example 2

A small smooth sphere is moving in the xy-plane and collides with a smooth fixed vertical wall which contains the y-axis. The velocity of the sphere just before impact is $(4\mathbf{j} - 5\mathbf{i})\,\mathrm{m\,s^{-1}}$. Given that the coefficient of restitution between the sphere and the wall is $\tfrac{1}{5}$,

(a) find the velocity of the sphere immediately after impact.
(b) Given further that the sphere is of mass 3 kg, find the loss of kinetic energy as a result of the impact.

This diagram summarises the given information:

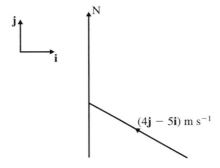

Let the velocity after the impact be $(u\mathbf{i} + v\mathbf{j})\,\mathrm{m\,s^{-1}}$. The components of velocity are then:

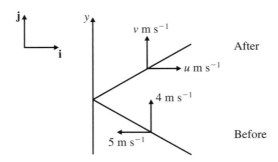

(i) *Parallel to the wall*
No change of velocity component $\Rightarrow v = 4$

(ii) *Perpendicular to the wall*
Newton's law of restitution $\Rightarrow u = \frac{1}{5} \times 5 = 1$

So the velocity of the sphere after the impact is $(\mathbf{i} + 4\mathbf{j})\,\mathrm{m\,s^{-1}}$.

(b) The kinetic energy of the sphere before impact is

$$\tfrac{1}{2} \times 3 \times (4^2 + 5^2) = \tfrac{3}{2} \times 41$$

The kinetic energy of the sphere after impact is

$$\tfrac{1}{2} \times 3 \times (1^2 + 4^2) = \tfrac{3}{2} \times 17$$

$$\text{Loss of K.E.} = \tfrac{3}{2}(41 - 17) = 36\,\mathrm{J}$$

Example 3

A smooth sphere strikes a smooth fixed wall with speed $u\,\mathrm{m\,s^{-1}}$ at an angle of $30°$ to the wall. It rebounds with speed $16\,\mathrm{m\,s^{-1}}$ at an angle of β to the wall. Given that the coefficient of restitution between the sphere and the wall is $\frac{1}{2}$, find u and β.

The given information is summarised in the figure:

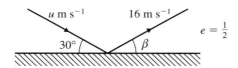

(i) *Parallel to the wall*

$$u\cos 30° = 16\cos\beta$$

$$\frac{u\sqrt{3}}{2} = 16\cos\beta \qquad\qquad (1)$$

(ii) *Perpendicular to the wall*
Newton's law of restitution gives:

$$16\sin\beta = \tfrac{1}{2}u\sin 30° = \tfrac{1}{2}u \times \tfrac{1}{2} = \tfrac{1}{4}u \qquad\qquad (2)$$

Squaring and adding equations (1) and (2) gives:

$$(16)^2\cos^2\beta + (16)^2\sin^2\beta = \tfrac{3}{4}u^2 + \tfrac{1}{16}u^2$$

or:

$$(16)^2 = u^2(\tfrac{3}{4} + \tfrac{1}{16}) = \tfrac{13}{16}u^2$$

$$\Rightarrow \qquad u^2 = \frac{(16)^2 \times 16}{13}$$

and $u = 17.8$

Dividing (2) by (1) gives:

$$\frac{16 \sin \beta}{16 \cos \beta} = \frac{\frac{1}{4}u}{u\frac{\sqrt{3}}{2}} = \frac{2}{4\sqrt{3}}$$

So:
$$\tan \beta = \frac{1}{2\sqrt{3}}$$

$\Rightarrow$
$$\beta = 16.1°$$

Example 4

A smooth sphere strikes a smooth fixed vertical wall with speed $u \, \text{m s}^{-1}$ at an angle of $60°$ to the wall. The sphere rebounds with speed $v \, \text{m s}^{-1}$ at an angle of $45°$ to the wall. Find the coefficient of restitution between the sphere and the wall.

The given information can be summarised in a diagram:

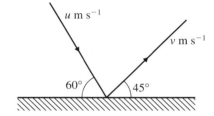

(i) *Parallel to the wall*

$$u \cos 60° = v \cos 45°$$

or
$$u \times \tfrac{1}{2} = v \times \frac{\sqrt{2}}{2} \tag{1}$$

(ii) *Perpendicular to the wall*
Newton's law of restitution gives:

$$v \sin 45° = eu \sin 60°$$

or
$$v\frac{\sqrt{2}}{2} = eu\frac{\sqrt{3}}{2} \tag{2}$$

Dividing (2) by (1) eliminates both u and v and gives:

$$e = \frac{1}{\sqrt{3}}$$

Exercise 2A

1 A smooth billiard ball is projected along a smooth horizontal table towards a fixed smooth vertical cushion. Before impact with the cushion it has a speed of $20 \, \text{m s}^{-1}$ and its direction of motion makes an angle of $30°$ with the cushion. Given that the

coefficient of restitution between the billiard ball and the cushion is $\frac{2}{5}$, find the magnitude and direction of the velocity of the billiard ball after impact.

2   A small smooth sphere is projected along a smooth horizontal table towards a fixed smooth vertical wall. Before impact with the wall it has a speed of $8\,\text{m s}^{-1}$ and its direction of motion makes an angle of $40°$ with the wall. Given that the coefficient of restitution between the sphere and the wall is $\frac{3}{10}$, find the magnitude and direction of the velocity of the sphere after impact.

3   A small smooth sphere of mass $2\,\text{kg}$ is moving in the xy-plane and collides with a smooth fixed vertical wall which contains the y-axis. The velocity of the sphere just before impact is $(2\mathbf{j} - 6\mathbf{i})\,\text{m s}^{-1}$. The coefficient of restitution between the sphere and the wall is $\frac{1}{2}$. Find:
(a) the velocity of the sphere after impact,
(b) the loss of kinetic energy due to the impact,
(c) the impulse exerted by the wall on the sphere.

4   A small smooth sphere of mass $1.5\,\text{kg}$ is moving in the xy-plane and collides with a smooth fixed vertical wall which contains the x-axis. The velocity of the sphere just before impact is $(2\mathbf{i} - 9\mathbf{j})\,\text{m s}^{-1}$. The coefficient of restitution between the sphere and the wall is $\frac{1}{3}$. Find:
(a) the velocity of the sphere after impact,
(b) the impulse exerted by the wall on the sphere.

5   A smooth billiard ball strikes a smooth cushion with speed $u\,\text{m s}^{-1}$ at an angle of $60°$ to the cushion. Given that the coefficient of restitution between the ball and the cushion is $\frac{1}{3}$, show that the ball rebounds at right angles to its original direction of motion.

6   A small smooth spherical ball of mass m falls vertically and strikes a fixed smooth inclined plane with speed u.
(a) Explain why the component of the velocity of the ball parallel to the plane is not affected by the impact.
The plane is inclined at $\alpha°$ to the horizontal, $\alpha < 45$. The ball rebounds horizontally.

(b) Show that $e = \tan^2 \alpha$.

(c) Show that a fraction $(1 - e)$ of the kinetic energy is lost during the impact.

(d) Show also that the magnitude of the impulse exerted on the sphere by the plane is $mu \sec \alpha$.

7   A smooth snooker ball strikes a smooth cushion when moving in a direction inclined at $60°$ to the cushion. The ball rebounds at an angle of $45°$ to the cushion. Show that one half of the kinetic energy of the ball is lost in the impact.

8   Two smooth vertical walls stand on a smooth horizontal floor and intersect at an acute angle θ. A small smooth particle is projected along the floor at right angles to one of the walls and away from it. After one impact with each wall the particle is moving parallel to the first wall it struck. Given that the coefficient of restitution between the particle and each wall is e show that:
$$(1 + 2e) \tan^2 \theta = e^2$$

9   Two smooth vertical walls are perpendicular to each other. A small smooth ball moving with speed $u \, \text{m s}^{-1}$ on a smooth horizontal surface strikes one wall. The initial direction of motion of the ball makes an angle α ($\alpha < 90°$) with the wall such that after the impact the ball moves towards the second wall. The coefficient of restitution between the ball and each wall is e. Show that after hitting the second wall the direction of motion of the ball is parallel to its original direction.

10   A smooth sphere S, moving on a smooth horizontal plane with speed $15 \, \text{m s}^{-1}$ strikes a smooth fixed vertical wall. The direction of motion of the ball before the impact makes an angle of $50°$ with the wall. After the impact S moves at an angle of $30°$ to the wall. Find:

(a) the coefficient of restitution between the sphere and the wall,

(b) the speed of S after the impact.

11   A particle P is moving with speed U on a smooth horizontal surface when it collides with a smooth vertical wall. The direction of motion of P is deflected through $90°$ by the impact and the coefficient of restitution between P and the wall is $\frac{2}{3}$.

(a) Find, to the nearest degree, the angle between P's original direction of motion and the wall.

(b) Show that the speed of P after the impact is $U\sqrt{\frac{2}{3}}$.

12 A small smooth marble is projected from a point on the edge of a smooth circular tray with a vertical side. The marble strikes the side of the tray twice and then returns to its starting point. The coefficient of restitution between the marble and the side of the tray is $\frac{1}{2}$. Find, to the nearest degree, the angle between the initial direction of motion of the ball and the radius at the point of projection.

2.2 Oblique impact of smooth elastic spheres

When you are studying the collision of two smooth elastic spheres, it is essential to consider separately:

(i) the components of the velocities perpendicular to the line of centres at impact,

(ii) the components of the velocities parallel to the line of centres at impact.

(i) Since there is no component of the mutual reaction in this direction, then, as before, **the components of the velocities perpendicular to the line of centres are unchanged**.

(ii) It is necessary to form two equations, one expressing **the conservation of momentum in this direction** and the other **Newton's law of restitution for components of velocity in this direction**. This exactly parallels the case of direct impact considered in chapter 4 of Book M2.

Consider two smooth spheres of mass m_1 and m_2, with centres C_1 and C_2, which collide, the coefficient of restitution between the two spheres being e. Just before impact suppose the speed of the first sphere is u_1 at angle α with the line of centres C_1C_2, and the speed of the second sphere is u_2 at angle β with the line of centres C_1C_2.

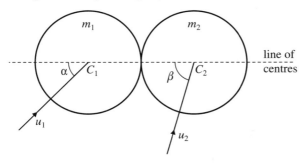

It is helpful to draw a diagram indicating the components of velocities of the spheres *before impact*, along and perpendicular to the line of centres C_1C_2.

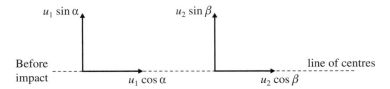

Then draw a diagram indicating the components of velocities of the spheres, *after impact*, along and perpendicular to the line of centres. The components of velocity perpendicular to C_1C_2, namely $u_1 \sin \alpha$ and $u_2 \sin \beta$, are unchanged by the impact. Suppose the components of the velocities along the line of centres after impact are v_1 and v_2 respectively.

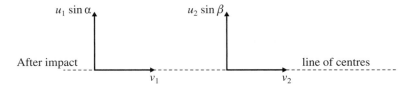

To obtain v_1 and v_2 you need two equations.

(i)   Total momentum along C_1C_2 is conserved:

$$m_1u_1 \cos \alpha + m_2u_2 \cos \beta = m_1v_1 + m_2v_2 \qquad (1)$$

(ii)   Newton's law of restitution:

$$\frac{\text{speed of separation along line of centres}}{\text{speed of approach along line of centres}} = e$$

$$\frac{v_2 - v_1}{u_1 \cos \alpha - u_2 \cos \beta} = e$$

or: $$v_2 - v_1 = e(u_1 \cos \alpha - u_2 \cos \beta) \qquad (2)$$

You can solve equations (1) and (2) to find v_1 and v_2. The resultant velocity of each of the spheres and their directions of motion can be found from v_1 and v_2 and the components of velocity perpendicular to C_1C_2, that is, $u_1 \sin \alpha$ and $u_2 \sin \beta$.

Example 5

A smooth sphere A, of mass 2 kg, and moving with speed $8 \, \mathrm{m \, s^{-1}}$ collides obliquely with a stationary sphere B, of mass 2 kg, the coefficient of restitution between the spheres being $\frac{1}{2}$. At the instant of impact the velocity of A makes an angle of 45° with the line of centres of the spheres. Find the magnitude and direction of the velocities of A and B immediately after impact.

The initial situation is summarised in the diagram:

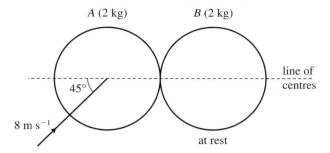

The components of velocity *before impact* are:

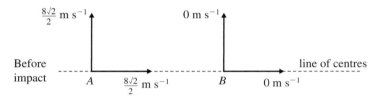

If the speeds of A and B along the line of centres immediately after impact are $v_1 \, \mathrm{m\,s^{-1}}$ and $v_2 \, \mathrm{m\,s^{-1}}$ then the components of velocity *after impact* are:

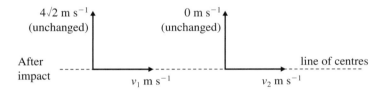

To obtain v_1 and v_2 write down the two usual equations:

(i) conservation of momentum gives:

$$2(4\sqrt{2} + 0) = 2v_1 + 2v_2$$

or
$$4\sqrt{2} = v_1 + v_2 \tag{1}$$

(ii) Newton's law of restitution gives:

$$v_2 - v_1 = \tfrac{1}{2}(4\sqrt{2} - 0) \tag{2}$$

Adding (1) and (2) gives:

$$2v_2 = 4\sqrt{2} + 2\sqrt{2}$$
$$v_2 = 3\sqrt{2} \tag{3}$$

Subtracting (2) from (1) gives:

$$2v_1 = 4\sqrt{2} - 2\sqrt{2}$$
$$v_1 = \sqrt{2} \tag{4}$$

For sphere A after impact the components of velocity are:

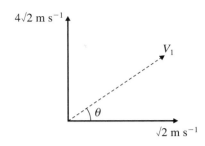

The magnitude of the speed V_1 is obtained from

$$V_1^2 = (4\sqrt{2})^2 + (\sqrt{2})^2 = 32 + 2 = 34$$

so:
$$V_1 = \sqrt{34}$$

The direction is obtained from:

$$\tan\theta = \frac{4\sqrt{2}}{\sqrt{2}} = 4$$

so:
$$\theta = 76.0°$$

So the speed of A is $\sqrt{34}\,\text{m s}^{-1}$, or $5.83\,\text{m s}^{-1}$, and the direction makes an angle of $76.0°$ with the line of centres.

For sphere B after impact the components of velocity are:

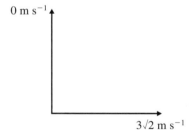

So B moves off along the line of centres with speed $3\sqrt{2}\,\text{m s}^{-1}$ or $4.24\,\text{m s}^{-1}$.

Example 6

A small smooth sphere A of mass 2 kg collides with a small smooth sphere B of mass 1 kg. The coefficient of restitution between the spheres is $\frac{1}{2}$. Just before impact A has a speed of $8\,\text{m s}^{-1}$ and is moving in a direction inclined at $30°$ to the line of centres and B has a speed of $4\,\text{m s}^{-1}$ and is moving in a direction inclined at $60°$ to the line of centres. Find the loss in kinetic energy as a result of the impact. Find also the magnitude of the impulse exerted by B on A.

Here is the situation before impact:

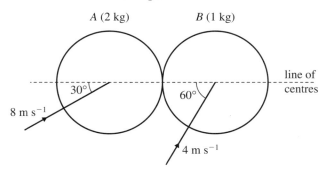

The components of velocity *before impact* are:

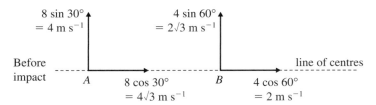

Let the speeds of A and B along the line of centres immediately after impact be $v_1 \text{ m s}^{-1}$ and $v_2 \text{ m s}^{-1}$. The components of velocity after impact are then:

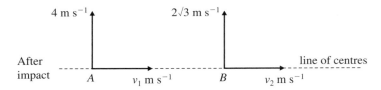

To obtain v_1 and v_2, write down the two usual equations:

(i) conservation of momentum:

$$2 \times 4\sqrt{3} + 1 \times 2 = 2v_1 + v_2$$

or: $\qquad\qquad 8\sqrt{3} + 2 = 2v_1 + v_2$ (1)

(ii) Newton's law of restitution:

$$v_2 - v_1 = \tfrac{1}{2}(4\sqrt{3} - 2)$$

or: $\qquad\qquad v_2 - v_1 = 2\sqrt{3} - 1$ (2)

Eliminating v_1 between (1) and (2) gives:

$$v_2 = 4\sqrt{3}$$ (3)

Subtracting (2) from (1) gives:

$$3v_1 = 6\sqrt{3} + 3$$

$$v_1 = 2\sqrt{3} + 1$$ (4)

The velocity components perpendicular to the line of centres are unchanged by the impact. So the loss of kinetic energy is solely due to change in velocities along the line of centres.

For sphere A,

$$\text{loss of K.E.} = \tfrac{1}{2} \times 2[(4\sqrt{3})^2 - (2\sqrt{3} + 1)^2]\,\text{J}$$
$$= [48 - 12 - 1 - 4\sqrt{3}]\,\text{J}$$
$$= (35 - 4\sqrt{3})\,\text{J}$$

For sphere B,

$$\text{loss of K.E.} = \tfrac{1}{2} \times 1[(2)^2 - (4\sqrt{3})^2]\,\text{J}$$
$$= \tfrac{1}{2}[4 - 48]\,\text{J}$$
$$= -22\,\text{J}$$

So total loss of K.E. is

$$(13 - 4\sqrt{3})\,\text{J} = 6.07\,\text{J}$$

For sphere A, change of momentum along line of centres

$$= 2 \times 4\sqrt{3} - 2(2\sqrt{3} + 1)$$
$$= 4\sqrt{3} - 2$$
$$= 4.93\,\text{Ns}$$

Hence the magnitude of the impulse exerted by B on A is $4.93\,\text{Ns}$.

Example 7

A smooth billiard ball A collides with a stationary identical ball B. The direction of motion of A before the impact makes an angle α with the line of centres at the moment of impact. The coefficient of restitution between the two balls is e $(e \neq 1)$. Show that ϕ, the angle through which the direction of motion of A is turned, satisfies

$$\tan \phi = \frac{(1 + e)\tan \alpha}{2\tan^2 \alpha + 1 - e}$$

Here is the situation just before impact:

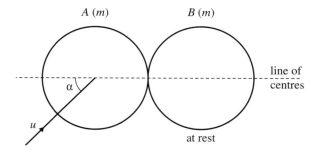

The components of velocity are:

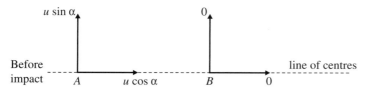

Let the speeds of A and B along the line of centres immediately after impact be v_1 and v_2. The components of velocity are then:

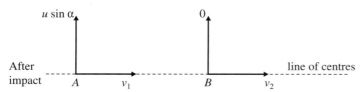

As before, you can get v_1 and v_2 by writing down the usual two equations.

(i)   Conservation of momentum:

$$mu \cos \alpha = mv_1 + mv_2$$

or: $$u \cos \alpha = v_1 + v_2 \qquad (1)$$

(ii)   Newton's law of restitution:

$$v_2 - v_1 = e(u \cos \alpha - 0)$$

or: $$v_2 - v_1 = eu \cos \alpha \qquad (2)$$

Subtracting (2) from (1) gives:

$$2v_1 = u \cos \alpha - eu \cos \alpha$$

so: $$v_1 = \tfrac{1}{2}u(1 - e) \cos \alpha$$

The components of the velocity of A after impact are:

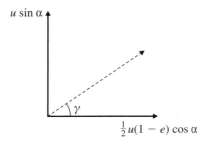

If the new direction of motion of A makes an angle γ with the line of centres, then:

$$\tan \gamma = \frac{u \sin \alpha}{\frac{1}{2}u(1 - e) \cos \alpha}$$

$$= \frac{2 \tan \alpha}{1 - e}$$

As may be seen from the diagram the angle through which the direction of motion of A has turned is

$$(\gamma - \alpha) = \phi$$

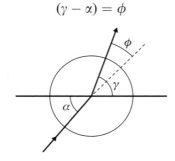

$$\tan \phi = \tan (\gamma - \alpha) = \frac{\tan \gamma - \tan \alpha}{1 + \tan \gamma \tan \alpha}$$

(Book P2 chapter 4)

$$= \frac{\dfrac{2 \tan \alpha}{1 - e} - \tan \alpha}{1 + \dfrac{2 \tan \alpha}{1 - e} \times \tan \alpha}$$

$$= \frac{(1 + e) \tan \alpha}{2 \tan^2 \alpha + 1 - e}$$

Example 8

Two identical smooth spheres are moving on a horizontal table with velocities $(3\mathbf{i} + 4\mathbf{j})\,\mathrm{m\,s}^{-1}$ and $(-\mathbf{i} + \mathbf{j})\,\mathrm{m\,s}^{-1}$. They collide when the line of centres is parallel to the vector $\mathbf{i}$. After impact the velocities are $4\mathbf{j}\,\mathrm{m\,s}^{-1}$ and $(2\mathbf{i} + \mathbf{j})\,\mathrm{m\,s}^{-1}$ respectively. Find the coefficient of restitution between the spheres.

As the information is given here in vector form, you can draw the diagram for velocity components immediately:

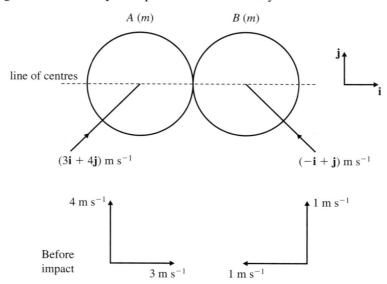

After impact the velocity components are:

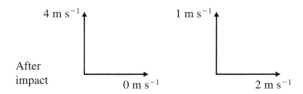

After
impact

Since you want to find e you need only write down the equation that expresses Newton's law of restitution. This gives:

$$(2 - 0) = e[3 - (-1)]$$

So:

$$2 = 4e$$

and:

$$e = \tfrac{1}{2}$$

Exercise 2B

1 A smooth sphere A of mass 2 kg and moving with speed
   $4 \, \text{m s}^{-1}$ collides with a stationary sphere B which has the same
   radius but a mass of 1 kg. The coefficient of restitution
   between the spheres is $\tfrac{1}{2}$. At the instant of impact the velocity
   of A makes an angle of $60°$ with the line of centres. Find the
   magnitude and direction of the velocities of A and B and the
   loss of kinetic energy as a result of the collision.

2 A small smooth sphere A of mass m collides with a stationary
   sphere B of the same radius but of mass M. At the instant of
   impact the velocity of A makes an angle of θ with the line of
   centres. The direction of motion of A is turned through a right
   angle by the impact. Show that

$$\tan^2 \theta = \frac{eM - m}{M + m}$$

   where e is the coefficient of restitution between the spheres.

3 Two identical smooth snooker balls, moving
   with equal speeds, collide as shown in the
   diagram. The coefficient of restitution
   between the balls is $\tfrac{1}{2}$. Find the speeds and
   direction of motion of the balls after the
   collision.

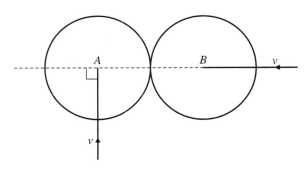

4 Two smooth spheres A and B have equal radii and mass 3 kg and 2 kg respectively. They are moving on a horizontal plane and collide. The coefficient of restitution between the spheres is $\frac{1}{4}$. At the moment of impact the line of centres is parallel to the unit vector $\mathbf{i}$. Immediately before impact the velocity of A is $(8\mathbf{i} + 4\mathbf{j})\,\mathrm{m\,s}^{-1}$ and the velocity of B is $(-4\mathbf{i} + 2\mathbf{j})\,\mathrm{m\,s}^{-1}$.

(a) Find the velocities of A and B immediately after impact.

(b) Find the angle between the velocities of A before and after impact.

(c) Find the loss of kinetic energy as a result of the impact.

5 A smooth billiard ball P, of mass m, moving with speed u, collides with an identical smooth billiard ball Q which is at rest. Just before impact the velocity of P makes an angle θ with the line of centres. Immediately after impact the velocity of P makes an angle ϕ with the line of centres. Given that the speeds of P and Q immediately after impact are equal, show that $\phi = 2\theta$. Deduce that e, the coefficient of restitution between the balls, is equal to $\tan^2 \theta$.

Find the speeds of P and Q after the impact and the loss of kinetic energy as a result of the impact in terms of m, u and e.

6 A smooth sphere A moving with speed u collides with an identical smooth sphere B at rest. Just before impact the speed of A makes an angle $\alpha\left(\alpha < \frac{\pi}{2}\right)$ with the line of centres. The coefficient of restitution between the spheres is $\frac{2}{3}$. Find the speeds of A and B after impact and show that if $\sin^2 \alpha = \frac{8}{35}$, the speed of A is halved by the impact.

7 A smooth sphere S of mass m is moving on a horizontal plane when it collides with another smooth sphere T of the same radius but of mass km $(k > 1)$ which is at rest. The sphere S strikes the sphere T obliquely. After the impact the two spheres are moving in perpendicular directions. Show that the coefficient of restitution is equal to $\dfrac{1}{k}$.

8 Two smooth spheres A and B have equal radii and masses m and $2m$ respectively. Sphere A is moving with velocity $a\mathbf{i} + a\mathbf{j}$ when it strikes sphere B, which is at rest. At the moment of impact the line of centres is parallel to the unit vector $\mathbf{i}$. After the impact the velocities of A and B are $v_1\mathbf{j}$ and $v_2\mathbf{i}$ respectively.

(a) Show that the coefficient of restitution is $\frac{1}{2}$.

(b) Find v_1 and v_2 in terms of a.

9 Two small smooth spheres of mass m_1 and m_2, with centres C_1 and C_2, collide obliquely, the coefficient of restitution between the two spheres being e. Just before impact the speed of the first sphere is u_1 at an angle α to C_1C_2 and the speed of the second sphere is u_2 at an angle β to C_1C_2. After the impact the components of the velocities along the line of centres are v_1 and v_2 respectively. Show that

$$v_1 = \frac{(m_1 - em_2)u_1 \cos\alpha + m_2u_2(1+e)\cos\beta}{(m_1 + m_2)}$$

$$v_2 = \frac{m_1u_1(1+e)\cos\alpha + (m_2 - em_1)u_2\cos\beta}{(m_1 + m_2)}$$

10 Two smooth spheres A and B have equal radii and masses m_1 and m_2 respectively. They are moving on a horizontal plane and collide. Just before impact the speed of A is u_1 and its direction of motion makes an angle α with the line of centres AB. The speed of B is u_2 and its direction of motion makes an angle β with the line of centres AB. Given that the coefficient of restitution between the spheres is e, show that the loss of kinetic energy as a result of the impact is

$$\frac{1}{2}\left(\frac{m_1m_2}{m_1 + m_2}\right)(u_1 \cos\alpha - u_2 \cos\beta)^2(1 - e^2)$$

11 A smooth ball of mass $2m$ is moving with speed U on a smooth horizontal plane. It collides with a second smooth ball of mass $5m$ which is at rest on the horizontal plane. The coefficient of restitution between the two balls is e. After the impact the directions of motion of the two balls are at right angles. By modelling the two balls as particles, find the value of e.

12 A small smooth ball A is moving with speed u on a smooth horizontal table when it hits an identical ball B which is at rest on the table. Before the impact the direction of motion of A makes an angle of $45°$ with the line of centres of the two balls at the moment of impact. The coefficient of restitution between the two balls is e.

(a) Find, in terms of u and e, the speed of B immediately after the impact.

(b) Find, in terms of u and e, the components of the velocity of A, parallel and perpendicular to the line of centres, immediately after the impact.

After the impact the direction of motion of A makes an angle of $\theta°$ with the line of centres. Given that $e = \frac{4}{5}$,

(c) find the value of θ.

13 A uniform small smooth sphere A, of mass m, moving with speed u on a smooth horizontal table collides with a stationary uniform small smooth sphere B of the same size as A and of mass m. The direction of motion of A before the impact makes an angle α with the line of centres of A and B, and the direction of motion of A after impact makes an angle β with the same line as shown in the figure below.

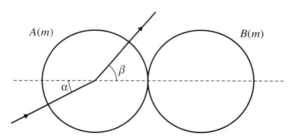

Given that the coefficient of restitution between the spheres is $\frac{1}{3}$:

(a) Show that $\tan \beta = 3 \tan \alpha$.

(b) Express $\tan (\beta - \alpha)$ in terms of t where $t = \tan \alpha$.

(c) Hence find, as α varies, the maximum angle of deflection of A caused by the impact.

14 A smooth sphere P of mass $4\,\text{kg}$ is moving on a smooth horizontal surface with velocity $(2\mathbf{i} + 3\mathbf{j})\,\text{m s}^{-1}$. A smooth sphere Q of mass $3\,\text{kg}$ and the same radius as P is moving on the same surface with velocity $(-2\mathbf{i} - \mathbf{j})\,\text{m s}^{-1}$. The spheres collide when their line of centres is parallel to $\mathbf{i}$. Given that the

coefficient of restitution between the spheres is $\frac{2}{3}$, calculate the loss in kinetic energy due to the impact.

SUMMARY OF KEY POINTS

1 **Impact of a smooth sphere on a fixed smooth plane**
   The component of the velocity of the sphere parallel to the plane is unchanged.

   The component of the velocity perpendicular to the plane after impact $= e \times$ the component of the velocity perpendicular to the plane before impact, where e is the coefficient of restitution between the sphere and the plane.

2 **Oblique impact of smooth elastic spheres**
   Perpendicular to line of centres:
   The components of the velocities in this direction are unchanged.

   Parallel to line of centres:
   (i) The linear momentum in this direction is conserved.
   (ii) Speed of separation along line of centres $= e \times$ speed of approach along line of centres (Newton's law of restitution), where e is the coefficient of restitution between the spheres.

Review exercise 1

1 [*In this question the velocities given are relative to the Earth. The unit vectors* **i** *and* **j** *are directed due East and due North respectively.*]

A helicopter pilot is informed that a lifeboat is drifting in the sea with velocity $(-2\mathbf{i} + 4\mathbf{j})\,\mathrm{km\,h^{-1}}$ and that, relative to his position, the lifeboat has position vector $5(2\mathbf{i} + \mathbf{j})\,\mathrm{km}$. The helicopter is travelling at its maximum speed of $100\,\mathrm{km\,h^{-1}}$ and the pilot immediately moves to intercept the lifeboat in the shortest possible time. Given that the constant velocity of the helicopter is $(u\mathbf{i} + v\mathbf{j})\,\mathrm{km\,h^{-1}}$ show that

(a) $u = 2v - 10$.

(b) $v^2 - 8v - 1980 = 0$.

(c) Calculate the time taken by the helicopter to reach the lifeboat giving your answer to the nearest one-tenth of a minute.

(d) Calculate the bearing of the course taken by the helicopter giving your answer to the nearest one-tenth of a degree. [E]

2 At time t two points P and Q have position vectors $\mathbf{p}$ and $\mathbf{q}$ respectively, where

$$\mathbf{p} = 2a\mathbf{i} + (a\cos\omega t)\mathbf{j} + (a\sin\omega t)\mathbf{k},$$

$$\mathbf{q} = (a\sin\omega t)\mathbf{i} - (a\cos\omega t)\mathbf{j} + 3a\mathbf{k}$$

and a, ω are constants. Find $\mathbf{r}$, the position vector of P relative to Q, and $\mathbf{v}$, the velocity of P relative to Q. Find also the values of t for which $\mathbf{r}$ and $\mathbf{v}$ are perpendicular. Determine the smallest and greatest distances between P and Q. [E]

3 A rugby player is running due North with speed $4\,\mathrm{m\,s^{-1}}$. He throws the ball horizontally and the ball has an initial velocity relative to the player of $6\,\mathrm{m\,s^{-1}}$ in the direction θ° West of South, i.e. on a bearing of $(180 + \theta)^\circ$, where $\tan\theta^\circ = \frac{4}{3}$. Find the magnitude and the direction of the initial velocity of the ball relative to a stationary spectator.

Find also the bearing on which the ball appears to move initially to the referee who is running with speed $2\sqrt{2}\,\mathrm{m\,s^{-1}}$ in a North-Westerly direction.

(Give all results to 3 significant figures, with bearings in degrees.) [E]

4 A river flows at a constant speed of $5\,\mathrm{m\,s^{-1}}$ between straight parallel banks which are 240 m apart. A boat crosses the river, travelling relative to the water at a constant speed of $12\,\mathrm{m\,s^{-1}}$. A man cycles at a constant speed of $4\,\mathrm{m\,s^{-1}}$ along the edge of one bank of the river in the direction opposite to the direction of flow of the river. At the instant when the boat leaves a point O on the opposite bank, the cyclist is 80 m downstream of O. The boat is steered relative to the water in a direction perpendicular to the banks. Taking $\mathbf{i}$ and $\mathbf{j}$ to be perpendicular horizontal unit vectors downstream and across the river from O respectively, express, in terms of $\mathbf{i}$ and $\mathbf{j}$, the velocities and the position vectors relative to O of the boat and the cyclist t seconds after the boat leaves O. Hence, or otherwise, calculate the time when the distance between the boat and the cyclist is least, giving this least distance. If, instead, the boat were to be steered so that it crosses the river from O to a point on the other bank directly opposite to O, show that this crossing would take approximately 22 seconds. [E]

5 The unit vectors $\mathbf{i}$ and $\mathbf{j}$ are directed due East and due North respectively. The airport B is due North of airport A. On a particular day the velocity of the wind is $(70\mathbf{i} + 25\mathbf{j})\,\mathrm{km\,h^{-1}}$. Relative to the air an aircraft flies with constant speed $250\,\mathrm{km\,h^{-1}}$. When the aircraft flies directly from A to B determine
(a) its speed, in $\mathrm{km\,h^{-1}}$ relative to the ground,
(b) the direction, to the nearest degree, in which it must head.
After flying from A to B, the aircraft returns directly to A.
(c) Calculate the ratio of the time taken on the outward flight to the time taken on the return flight. [E]

6 At time $t = 0$, a particle A, which moves with constant velocity $(3\mathbf{i} + u\mathbf{j} + 5\mathbf{k})\,\mathrm{m\,s^{-1}}$, has position vector $(4\mathbf{i} + 9\mathbf{j} - 10\mathbf{k})\,\mathrm{m}$ relative to a fixed origin O. At time $t = 0$, a particle B, which has constant velocity $2\mathbf{i}\,\mathrm{m\,s^{-1}}$, is at O.

(a) Given that $u = 2$, find the value of t at the instant when the distance between A and B is least.

(b) At time $t\,\mathrm{s}$, a third particle C has position vector

$$[10\mathbf{i} + 5\mathbf{j}\sin(\pi t/4) + 5\mathbf{k}\cos(\pi t/4)]\,\mathrm{m}$$

relative to the fixed origin O. Find the value of u in this case, given that A and C collide.   [E]

7 At time t a particle P, of mass m, has position vector $\mathbf{p}$, given by

$$\mathbf{p} = (3a\cos\omega t)\mathbf{i} + (4a\sin\omega t)\mathbf{j}$$

where a and ω are positive constants. Find, in terms of m, a, ω and t

(a) the kinetic energy of P,

(b) the magnitude of the force acting on P.

A second particle Q has position vector $\mathbf{q}$, given by

$$\mathbf{q} = (3a\sin\omega t)\mathbf{j} + (4a\cos\omega t)\mathbf{k}$$

Find $\mathbf{r}$, the position vector of P relative to Q.

Evaluate $\mathbf{r} \cdot \mathbf{r}$ and hence, or otherwise, show that the greatest and least distances between P and Q are $5a$ and a respectively.   [E]

8 Two cyclists, C and D, are travelling with constant velocities $(5\mathbf{i} - 2\mathbf{j})\,\mathrm{m\,s^{-1}}$ and $8\mathbf{j}\,\mathrm{m\,s^{-1}}$ respectively relative to a fixed origin O.

(a) Find the velocity of C relative to D.

At noon, the position vectors of C and D are $(100\mathbf{i} + 300\mathbf{j})$ metres and $(150\mathbf{i} + 100\mathbf{j})$ metres respectively, referred to O. At t seconds after noon, the position vector of C relative to D is $\mathbf{s}$ metres.

(b) Show that $\mathbf{s} = (-50 + 5t)\mathbf{i} + (200 - 10t)\mathbf{j}$.

(c) By considering $|\mathbf{s}|^2$, or otherwise, find the value of t for which C and D are closest together.   [E]

9 Two joggers, A and B are each running with constant velocity on level parkland. At a certain instant, A and B have position vectors $(-60\mathbf{i} + 210\mathbf{j})\,\mathrm{m}$ and $(30\mathbf{i} - 60\mathbf{j})\,\mathrm{m}$ respectively, referred to a fixed origin O. Ninety seconds later, A and B meet at the point with position vector $(210\mathbf{i} + 120\mathbf{j})\,\mathrm{m}$.
(a) Find, as a vector in terms of $\mathbf{i}$ and $\mathbf{j}$, the velocity of A relative to B.
(b) Verify that the magnitude of the velocity of A relative to B is equal to the speed of A. [E]

10 [*In this question the velocities given are relative to the Earth. The unit vectors $\mathbf{i}$ and $\mathbf{j}$ are directed due East and due North respectively.*]
At time $t = 0$, two ice skaters P and Q have position vectors $2\mathbf{j}$ metres and $2\sqrt{3}\mathbf{i}$ metres respectively, referred to an origin O at the centre of the ice rink. The velocity of P is constant and equal to $3\mathbf{j}\,\mathrm{m\,s}^{-1}$ and the velocity of Q is constant and equal to $\mathbf{v}\,\mathrm{m\,s}^{-1}$. Skater Q moves in a straight line and at time $t = T$ seconds collides with P.
(a) Give a reason why $-2\sqrt{3}\mathbf{i} + 2\mathbf{j}$ is a vector in the direction of the velocity of Q relative to P.
(b) Show that

$$\mathbf{v} = -K\sqrt{3}\mathbf{i} + (K + 3)\mathbf{j},$$

where K is a positive constant.
Given that the speed of Q is $3\sqrt{3}\,\mathrm{m\,s}^{-1}$,
(c) find the value of K,
(d) find the value of T. [E]

11 Two straight roads cross at right angles at O, one road running North–South and the other East–West. At noon, A leaves the crossroads and walks due East at a speed of $6\,\mathrm{km\,h}^{-1}$ and at the same time B leaves a point $5\,\mathrm{km}$ due North of the crossroads and walks due South at $8\,\mathrm{km\,h}^{-1}$. Calculate:
(a) the magnitude and the direction of the velocity of B relative to A,
(b) the time at which A and B are closest together,
(c) the distance between A and B at this instant. [E]

12 A cyclist is moving due North at a constant speed of $5\,\mathrm{m\,s^{-1}}$. The wind is blowing at a constant speed of $3\,\mathrm{m\,s^{-1}}$ from the direction bearing $060°$. Calculate the magnitude of the velocity of the wind relative to the cyclist. [E]

13 A ship is steaming at 15 knots due East, while the wind speed is 20 knots from due North. Find the magnitude and the direction, to the nearest degree, of the wind velocity relative to the ship. Find also the course, between East and South, along which the ship would have to steer at 16 knots for the wind velocity relative to the ship to be at right angles to the course of the ship. Obtain the magnitude of the velocity of the wind relative to the ship in this case. [E]

14 A cyclist A is travelling with a constant velocity of $10\,\mathrm{km\,h^{-1}}$ due East and a cyclist B has a constant velocity of $8\,\mathrm{km\,h^{-1}}$ in a direction $\arctan(4/3)$ East of North. At noon, B is $0.6\,\mathrm{km}$ due South of A. Taking the position of B at noon as the origin and $\mathbf{i}$ and $\mathbf{j}$ as unit vectors due East and due North respectively, obtain expressions for the position vectors of A and B at time t hours after noon.
Hence find, at time t hours after noon,
(a) the position vector of A relative to B,
(b) the velocity of A relative to B.
Show that the cyclists are nearest together at time 4.8 minutes after noon and that, at this time, the distance between them is $360\,\mathrm{m}$. [E]

15 Fishing boat A is moving with constant velocity of magnitude $10\,\mathrm{m\,s^{-1}}$ in the direction $060°$. Fishing boat B is moving with constant velocity of magnitude $16\,\mathrm{m\,s^{-1}}$ due North. Show that the velocity of A relative to B is of magnitude $14\,\mathrm{m\,s^{-1}}$ and calculate the direction of this velocity, giving your answer as a bearing to the nearest tenth of a degree.
At noon, A is $5000\,\mathrm{m}$ due West of B and T seconds later the distance between A and B is least. Calculate the value of T to the nearest integer and determine the least distance between A and B, giving your answer to the nearest ten metres.
Given that visibility is limited to $5000\,\mathrm{m}$, state the time, in seconds, for which the boats remain in visual contact. [E]

16 A ship A is moving at a constant velocity of $16 \, \text{km h}^{-1}$ on a bearing of $060°$. At 10 a.m. ship B is 10 km from A, the bearing of A from B being $330°$. B is moving at a constant speed of $12 \, \text{km h}^{-1}$. Find the course B must set in order to pass as close as possible to A, giving your answer as a bearing.

17 Two equal smooth spheres approach each other from opposite directions with equal speeds. The coefficient of restitution between the spheres is e. At the moment of impact, their common normal is inclined at an angle θ to the original direction of motion. Given that after impact each sphere moves at right angles to its original direction of motion, prove that

$$\tan \theta = \sqrt{e}$$

[E]

18 A smooth uniform sphere A, of mass $2m$, is at rest on a smooth horizontal table. A second smooth sphere B, of mass m and the same radius, is moving with speed u and collides with the first sphere. At impact the direction of motion of sphere B makes an angle of $45°$ with the line of centres of the two spheres. The coefficient of restitution is $\frac{1}{2}$. Show that after impact the direction of motion of sphere B is perpendicular to the line of centres. Show also that the loss of kinetic energy due to the collision is $\frac{1}{8} mu^2$.

[E]

19 A smooth sphere A, of mass m and radius a, moves on a horizontal table with speed u and collides with a smooth stationary sphere B, of mass λm ($\lambda > 1$) and radius a. Before impact the direction of motion of A makes an acute angle θ with the line of centres of the spheres. As a result of the impact A is deflected through a right angle. The coefficient of restitution between the spheres is e.

Prove that $e > \dfrac{1}{\lambda}$ and show that

$$\tan^2 \theta \leqslant \frac{\lambda - 1}{\lambda + 1}$$

Given that $\lambda = 5$ and $\theta = \dfrac{\pi}{6}$, find e and the kinetic energy lost as a result of the impact.

[E]

20 Two smooth billiard balls A and B each of radius a have mass m and M respectively. B is initially at rest and A approaches it with speed u. The perpendicular distance of the centre of B from the line of motion of the centre of A is d. The coefficient of restitution between the spheres is e. Show that the kinetic energy lost due to the impact is

$$\frac{mMu^2}{2(m+M)}(1-e^2)\left(1-\frac{d^2}{4a^2}\right)$$ [E]

21 A smooth circular horizontal table is surrounded by a smooth rim whose interior surface is vertical. AB is a diameter of the circle.

Two equal particles of mass m are projected simultaneously from A each with speed V in directions making angles of $30°$ with AB, one on each side of the diameter. After only one impact each at the rim, the particles meet at a point P on AB where $AP < AB$.

(a) Show that the coefficient of restitution e between each particle and the rim satisfies $e > \frac{1}{3}$.

When the particles meet, they coalesce.

(b) Find the speed of the combined particle, in terms of e and V.

(c) Given that the kinetic energy of the combined particle is $\frac{1}{16}mV^2$, find the value of e. [E]

22 Two smooth spheres A and B of equal radius, but of mass $3m$ and $2m$ respectively, moving on a smooth horizontal table collide. At the moment of impact the line joining their centres is parallel to the unit vector $\mathbf{i}$. The unit vector $\mathbf{j}$ is in the plane of the table and is perpendicular to $\mathbf{i}$. Immediately before collision the velocities of A and B are $u(4\mathbf{i}+2\mathbf{j})$ and $u(-2\mathbf{i}+\mathbf{j})$ respectively, where u is a positive constant. Given that the coefficient of restitution between the spheres is $\frac{1}{4}$, find

(a) the velocities of A and B immediately after impact

(b) the cosine of the angle between the velocities of A before and after impact

(c) the kinetic energy lost as a result of the impact.

After impact, sphere B receives an impulsive blow which brings it to rest. Find the impulse of the blow. [E]

23 A smooth sphere A collides with an identical stationary sphere B so that the angle between the velocity of A and the line of centres of the spheres is α immediately before the collision and β immediately afterwards. Show that
$$2 \tan \alpha = (1 - e) \tan \beta$$
where e is the coefficient of restitution.

Show also that, as α varies but e remains fixed, the maximum deflection that can be produced in the direction of motion of A is δ, where
$$\sqrt{(8 - 8e)} \tan \delta = (1 + e) \qquad \text{[E]}$$

24 A smooth sphere A rests on a smooth horizontal floor at a distance d from a plane vertical wall. An identical sphere B is projected along the floor with speed V towards the wall in a direction perpendicular to the wall. Before striking the wall, sphere B strikes sphere A so that the line of centres on impact is inclined at an angle α to the wall. The coefficient of restitution between the spheres is $\frac{2}{3}$. Show that the components of the velocity of B after impact, along and perpendicular to the line of centres, are $\dfrac{V \sin \alpha}{6}$ and $V \cos \alpha$.

Given that the diameter of either sphere is negligible compared to d, show also that the distance between the points where the spheres strike the wall is
$$\frac{6d \cot \alpha}{1 + 5 \cos^2 \alpha} \qquad \text{[E]}$$

25 Two identical, small, smooth uniform spheres, P and Q, are travelling towards each other in parallel but opposite directions on a smooth horizontal table. The spheres collide and, at the instant of collision, their line of centres makes an acute angle θ with their original lines of motion. The coefficient of restitution between P and Q is e and the speeds of P and Q immediately before the collision are $2u$ and u respectively.

After the collision, Q moves in a direction at right angles to its original line of motion.

(a) Show that
$$\tan \theta = \sqrt{\left(\frac{1 + 3e}{2} \right)}.$$

(b) Deduce the range of possible values for $\tan \theta$. $\qquad$ [E]

26 A small smooth sphere T is at rest on a smooth horizontal table. An identical sphere S moving on the table with speed U collides with T. The directions of motion of S before and after impact make angles of $30°$ and $\beta°$ $(0 < \beta < 90)$ respectively with L, the line of centres at the moment of impact. The coefficient of restitution between S and T is e.

(a) Show that V, the speed of T immediately after impact, is given by

$$V = \frac{U\sqrt{3}}{4}(1 + e).$$

(b) Find the components of the velocity of S, parallel and perpendicular to L, immediately after impact.

Given that $e = \frac{2}{3}$,

(c) find, to 1 decimal place, the value of β.   [E]

27

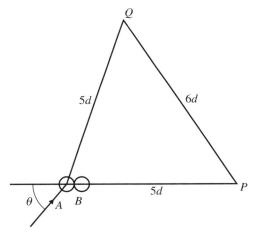

A small smooth sphere A moving with speed u on a horizontal table collides with another identical sphere B which is at rest on the table. The direction of motion of A before impact makes an angle θ, where $\tan \theta = \frac{3}{2}$, with the line of centres of A and B. The coefficient of restitution between the two spheres is $\frac{1}{8}$. Two points P and Q on the table are each at a distance $5d$ from the position of the centre of A at the moment of impact and the distance PQ is $6d$ as shown in the diagram. After the impact B moves towards P.

(a) Find the velocity of A after impact.

(b) Show that A moves towards Q.   [E]

Further motion of particles in one dimension

3

In Books M1, M2 and M3 you studied the kinematics of a particle moving in a straight line. In Book M1 the case of constant acceleration was dealt with. This was generalised in Book M2 to situations where the acceleration is a function of time (t) and in M3 to situations where the acceleration is a function of the displacement (x). Now we are going to consider the case where the acceleration is a function of the speed (v).

In much of the modelling dealt with in Books M1, M2 and M3 we assumed that air resistance could be neglected. But air resistance in many circumstances is *not* negligible and is not constant. In fact the magnitude of the resistance depends on the speed of the particle and is usually taken to be proportional to the speed for low speeds and proportional to the square of the speed for higher speeds. So when air resistance is taken into account the acceleration will be a function of the speed. We will consider only circumstances in which the acceleration is either

$$a + bv, \text{ where } a \text{ and } b \text{ are constants}$$

or

$$\alpha + \beta v^2, \text{ where } \alpha \text{ and } \beta \text{ are constants.}$$

3.1 Kinematics of a particle moving in a straight line when the acceleration is a function of speed

Consider a particle P moving in a straight line. Suppose that at time t seconds the displacement of P from a fixed point O in the line is x metres, its velocity is $v\,\mathrm{m\,s^{-1}}$ and its acceleration is $a\,\mathrm{m\,s^{-2}}$.

You should recall from chapter 1 of Book M3 that

(i) $$a = \frac{\mathrm{d}v}{\mathrm{d}t}$$

and (ii) $$a = v\frac{\mathrm{d}v}{\mathrm{d}x} = \frac{\mathrm{d}}{\mathrm{d}x}(\tfrac{1}{2}v^2)$$

When a is a function of v:

$$a = f(v)$$

so, using (i):

$$\frac{dv}{dt} = f(v)$$

Integrating:

$$\int \frac{dv}{f(v)} = \int dt$$

or

$$t = \int \frac{dv}{f(v)} + c \qquad (1)$$

where c is an arbitrary constant of integration.

Alternatively, using (ii):

$$v\frac{dv}{dx} = f(v)$$

Integrating:

$$\int \frac{v\,dv}{f(v)} = \int dx$$

or

$$x = \int \frac{v\,dv}{f(v)} + k \qquad (2)$$

where k is an arbitrary constant of integration.

When you do the integration over v in equation (2), you get:

$$x = F(v) + k \qquad (3)$$

In some cases you may be able, by writing $v = \dfrac{dx}{dt}$, to integrate (3) to obtain a relationship between x and t.

Example 1

A particle P is moving along Ox with an acceleration of $-(9 + v^2)\,\mathrm{m\,s^{-2}}$ at time t seconds when its displacement from O is x metres and its speed is $v\,\mathrm{m\,s^{-1}}$. When $t = 0$, $x = 0$ and $v = u$. Find the value of t and the value of x when the particle comes to rest.

Using:

$$a = \frac{dv}{dt} = -(9 + v^2)$$

gives:

$$\int \frac{dv}{9 + v^2} = -t + c$$

Integrating:

$$\tfrac{1}{3}\arctan\left(\frac{v}{3}\right) = -t + c$$

(See Book P5, chapter 3.)

Substituting $v = u$ when $t = 0$ gives:

$$c = \tfrac{1}{3}\arctan\left(\frac{u}{3}\right)$$

So:
$$t = \tfrac{1}{3}\arctan\left(\frac{u}{3}\right) - \tfrac{1}{3}\arctan\left(\frac{v}{3}\right)$$

The particle comes to rest when $v = 0$. Then:
$$t = \tfrac{1}{3}\arctan\left(\frac{u}{3}\right)$$

Using:
$$a = v\frac{dv}{dx} = -(9 + v^2)$$

gives:
$$\int \frac{v\,dv}{(9 + v^2)} = -x + k$$

Integrating:
$$\tfrac{1}{2}\ln(9 + v^2) = -x + k$$

Substituting $v = u$ when $x = 0$ gives
$$k = \tfrac{1}{2}\ln(9 + u^2)$$

So:
$$x = \tfrac{1}{2}\ln(9 + u^2) - \tfrac{1}{2}\ln(9 + v^2)$$

The particle comes to rest when $v = 0$.

So:
$$x = \tfrac{1}{2}\ln(9 + u^2) - \tfrac{1}{2}\ln 9$$
$$= \tfrac{1}{2}\ln\left(1 + \frac{u^2}{9}\right)$$

Example 2

A particle P is moving along Ox with an acceleration of $-(5 + 2v)\,\mathrm{m\,s^{-2}}$ at time t seconds when its displacement from O is x metres and its speed is $v\,\mathrm{m\,s^{-1}}$. When $t = 0$, $x = 0$ and $v = 10$. Find the value of t and the value of x when $v = 5$.

Using:
$$a = \frac{dv}{dt} = -(5 + 2v)$$

gives:
$$\int \frac{dv}{5 + 2v} = -t + c$$

Integrating:
$$\tfrac{1}{2}\ln(5 + 2v) = -t + c$$

Substituting $v = 10$ when $t = 0$ gives:
$$c = \tfrac{1}{2}\ln(25)$$

So:
$$t = \tfrac{1}{2}\ln(25) - \tfrac{1}{2}\ln(5 + 2v)$$

When $v = 5$,
$$t = \tfrac{1}{2}\ln 25 - \tfrac{1}{2}\ln(5 + 10)$$
$$= \tfrac{1}{2}\ln\left(\tfrac{25}{15}\right)$$
$$= \tfrac{1}{2}\ln\left(\tfrac{5}{3}\right) = 0.255$$

Using:
$$a = v\frac{dv}{dx} = -(5+2v)$$

gives:
$$\int \frac{v\,dv}{5+2v} = -x + k$$

To integrate the left-hand side, divide out to obtain:

$$\int \left[\tfrac{1}{2} - \frac{\frac{5}{2}}{(5+2v)} \right] dv$$

Integrating:
$$\tfrac{1}{2}v - \tfrac{5}{2} \times \tfrac{1}{2}\ln(5+2v) = -x + k$$

Substituting $v = 10$ when $x = 0$ gives:

$$k = \tfrac{1}{2}(10) - \tfrac{5}{4}\ln(5 + 2 \times 10)$$
$$= 5 - \tfrac{5}{4}\ln(25)$$

So:
$$x = 5 - \tfrac{5}{4}\ln(25) - \tfrac{1}{2}v + \tfrac{5}{4}\ln(5+2v)$$

When $v = 5$,

$$x = 5 - \tfrac{5}{4}\ln(25) - \tfrac{1}{2}(5) + \tfrac{5}{4}\ln(15)$$
$$= 2\tfrac{1}{2} - \tfrac{5}{4}\ln\left(\tfrac{5}{3}\right)$$
$$= 1.86$$

So when $v = 5$, $t = 0.255\,\text{s}$ and $x = 1.86\,\text{m}$.

3.2 Vertical motion taking into account air resistance

Resistance proportional to speed

Consider a particle P of mass m and suppose the air resistance is of magnitude mkv, where v is the speed and k is a positive constant.

(i) Particle falling from rest at a point O

Measure the displacement *downwards* from O, because O is a **fixed point**.

■ **Displacements must always be measured from a fixed point.**

The forces acting on the particle are its weight mg and the resistance, as shown in the diagram. The equation of motion is

$$m \times \text{acceleration} = mg - mkv$$

So:
$$\text{acceleration} = a = g - kv$$

Using $a = \dfrac{dv}{dt}$ gives:

$$\frac{dv}{dt} = g - kv \qquad (1)$$

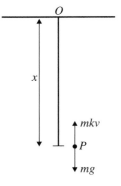

Integrating:
$$\int \frac{dv}{g - kv} = t + c$$

So:
$$-\frac{1}{k}\ln(g - kv) = t + c$$

Substituting $v = 0$ when $t = 0$ gives:
$$c = -\frac{1}{k}\ln g$$

and
$$t = \frac{1}{k}\ln g - \frac{1}{k}\ln(g - kv) = \frac{1}{k}\ln\left(\frac{g}{g - kv}\right)$$

Hence:
$$e^{kt} = \frac{g}{g - kv}$$

or:
$$(g - kv) = ge^{-kt}$$

so that:
$$v = \frac{g}{k}(1 - e^{-kt}) \qquad (2)$$

Since
$$\lim_{t \to \infty}(1 - e^{-kt}) = 1$$

the speed v tends to $\frac{g}{k}$. This is called the **terminal velocity**.

If you substitute this value in the differential equation you get:
$$a = \frac{dv}{dt} = g - kv$$
$$= g - k\frac{g}{k}$$
$$= g - g$$
$$= 0$$

So if and when v reached the value $\frac{g}{k}$ the acceleration would become zero and the particle would then descend with *uniform* speed. This is a plot of v against t:

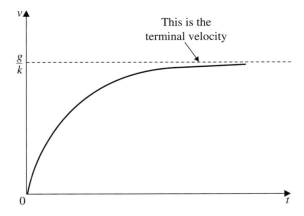

Using $a = v\dfrac{\mathrm{d}v}{\mathrm{d}x}$ gives:

$$v\frac{\mathrm{d}v}{\mathrm{d}x} = g - kv$$

Integrating:

$$\int \frac{v\,\mathrm{d}v}{g - kv} = x + A$$

To carry out the integration over v we write

$$v = -\frac{1}{k}(g - kv) + \frac{g}{k}$$

as $(g - kv)$ occurs in the denominator. Using this gives

$$\int \left(-\frac{1}{k} + \frac{g}{k}\frac{1}{(g - kv)}\right)\mathrm{d}v = x + A$$

Integrating over v gives

$$-\frac{1}{k}v - \frac{g}{k^2}\ln(g - kv) = x + A$$

But $v = 0$ when $x = 0$.

So:

$$A = -\frac{g}{k^2}\ln g$$

and

$$x = \frac{g}{k^2}\ln\left(\frac{g}{g - kv}\right) - \frac{v}{k}$$

In this case you can integrate equation (2) (on page 59) to obtain x as a function of t.

From (2):

$$v = \frac{\mathrm{d}x}{\mathrm{d}t} = \frac{g}{k}(1 - \mathrm{e}^{-kt})$$

Integrating:

$$x = \frac{gt}{k} + \frac{g}{k^2}\mathrm{e}^{-kt} + C$$

But $x = 0$ when $t = 0$

So:

$$C = -\frac{g}{k^2}$$

and

$$x = \frac{g}{k^2}(kt + \mathrm{e}^{-kt} - 1)$$

(ii) Particle projected vertically upwards with speed U

This time, measure the displacement *upwards* from the point of projection O', since O' is a fixed point. The forces acting on the particle are shown in the diagram. If the speed of the particle is u, then the equation of motion is

$$a = -g - ku$$

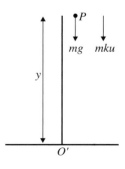

Using $a = \dfrac{du}{dt}$ gives:

$$\frac{du}{dt} = -g - ku$$

Integrating:

$$\int \frac{du}{g + ku} = -t + c$$

So:

$$\frac{1}{k} \ln (g + ku) = -t + c$$

Substituting $u = U$ when $t = 0$ gives:

$$c = \frac{1}{k} \ln (g + kU)$$

so that

$$\frac{1}{k} \ln (g + ku) - \frac{1}{k} \ln (g + kU) = -t$$

and

$$t = \frac{1}{k} \ln \left(\frac{g + kU}{g + ku} \right)$$

You can find the time the particle takes to reach the highest point of the path by setting $u = 0$.

So:

$$\text{time} = \frac{1}{k} \ln \left(1 + \frac{kU}{g} \right)$$

Using $a = u \dfrac{du}{dy}$ gives:

$$u \frac{du}{dy} = -g - ku$$

Integrating:

$$\int \frac{u \, du}{g + ku} = -y + A$$

or:

$$\frac{1}{k} \int \left(1 - \frac{g}{g + ku} \right) du = -y + A$$

So:

$$\frac{1}{k} u - \frac{g}{k^2} \ln (g + ku) = -y + A$$

Substituting $u = U$ when $y = 0$ gives:

$$A = \frac{1}{k} U - \frac{g}{k^2} \ln (g + kU)$$

So:

$$y = \frac{1}{k} (U - u) + \frac{g}{k^2} \ln \left(\frac{g + ku}{g + kU} \right)$$

You can find the greatest height h that the particle reaches by setting $u = 0$. Thus:

$$h = \frac{1}{k} U + \frac{g}{k^2} \ln \left(\frac{g}{g + kU} \right)$$

Resistance proportional to the square of the speed

Consider a particle P of mass m subject to a resistance of magnitude mkv^2, where v is the speed and k is a positive constant.

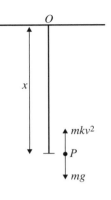

(i) Particle falling from rest at a point O

As before, measure the displacement downwards from O. The forces now acting on the particle are shown in the diagram. The equation of motion is:

$$m \times \text{acceleration} = mg - mkv^2$$

First notice that if and when v reaches the value $\sqrt{\left(\dfrac{g}{k}\right)}$ the acceleration would become zero and P would continue to descend with constant speed $\sqrt{\left(\dfrac{g}{k}\right)}$. In what follows, therefore, you can assume that $v < \sqrt{\left(\dfrac{g}{k}\right)}$.

Using $a = \dfrac{\mathrm{d}v}{\mathrm{d}t}$ gives:

$$\frac{\mathrm{d}v}{\mathrm{d}t} = g - kv^2$$

Integrating:

$$\int \frac{\mathrm{d}v}{g - kv^2} = t + c$$

You can write the left-hand side of this equation as partial fractions (see Book P3, chapter 1). It becomes

$$\frac{1}{k} \times \frac{1}{2\sqrt{\left(\dfrac{g}{k}\right)}} \int \left[\frac{1}{\sqrt{\left(\dfrac{g}{k}\right)} + v} + \frac{1}{\sqrt{\left(\dfrac{g}{k}\right)} - v} \right] \mathrm{d}v$$

$$= \frac{1}{2\sqrt{(kg)}} \ln \left[\frac{\sqrt{\left(\dfrac{g}{k}\right)} + v}{\sqrt{\left(\dfrac{g}{k}\right)} - v} \right]$$

So:

$$\frac{1}{2\sqrt{(kg)}} \ln \left[\frac{\sqrt{\left(\dfrac{g}{k}\right)} + v}{\sqrt{\left(\dfrac{g}{k}\right)} - v} \right] = t + c$$

Substituting $v = 0$ when $t = 0$ gives $c = 0$, and so:

$$\frac{1}{2\sqrt{(kg)}} \ln \left(\frac{\sqrt{\left(\dfrac{g}{k}\right)} + v}{\sqrt{\left(\dfrac{g}{k}\right)} - v} \right) = t$$

or:

$$\left(\frac{\sqrt{\left(\frac{g}{k}\right)} + v}{\sqrt{\left(\frac{g}{k}\right)} - v}\right) = e^{2t\sqrt{(kg)}}$$

so that:

$$e^{2t\sqrt{(kg)}}\left(\sqrt{\left(\frac{g}{k}\right)} - v\right) = \sqrt{\left(\frac{g}{k}\right)} + v$$

Solving for v:

$$v = \sqrt{\left(\frac{g}{k}\right)}\left[\frac{e^{2t\sqrt{(kg)}} - 1}{e^{2t\sqrt{(kg)}} + 1}\right]$$

or:

$$v = \sqrt{\left(\frac{g}{k}\right)}\left[\frac{1 - e^{-2t\sqrt{(kg)}}}{1 + e^{-2t\sqrt{(kg)}}}\right]$$

As $t \to \infty$ the speed $v \to \sqrt{\left(\frac{g}{k}\right)}$ which is called the **terminal speed** of the particle, or sometimes, rather loosely, the terminal velocity. Here is a plot of v against t:

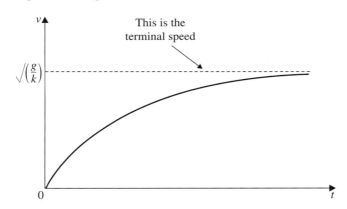

Using $a = v\dfrac{dv}{dx}$ gives:

$$v\frac{dv}{dx} = g - kv^2$$

Integrating:

$$\int \frac{v\, dv}{g - kv^2} = x + A$$

$$-\frac{1}{2k}\ln\left(g - kv^2\right) = x + A$$

$\left(\text{No modulus sign is required because } v < \sqrt{\left(\frac{g}{k}\right)}.\right)$

Substituting $x = 0$ when $v = 0$ gives:

$$A = -\frac{1}{2k}\ln g$$

and so:
$$x = \frac{1}{2k} \ln\left(\frac{g}{g - kv^2}\right)$$

or
$$v^2 = \frac{g}{k}(1 - e^{-2kx})$$

(ii) Particle projected vertically upwards with speed U

As before, measure the displacement *upwards* from the point of projection O'. The forces acting on the particle are shown in the diagram.

The speed of the particle is $u\,\mathrm{m\,s}^{-1}$ and the equation of motion is:

$$a = -g - ku^2$$

Using $a = \dfrac{du}{dt}$ gives:

$$\frac{du}{dt} = -g - ku^2$$

Integrating:
$$\int \frac{du}{g + ku^2} = -t + c$$

or
$$\frac{1}{k} \int \frac{du}{\left(\frac{g}{k}\right) + u^2} = -t + c$$

$\Rightarrow$
$$\frac{1}{\sqrt{(kg)}} \arctan\left[u\sqrt{\left(\frac{k}{g}\right)}\right] = -t + c$$

Since $u = U$ when $t = 0$:

$$c = \frac{1}{\sqrt{(kg)}} \arctan\left[U\sqrt{\left(\frac{k}{g}\right)}\right]$$

So:
$$\sqrt{(kg)}\,t = \arctan\left[U\sqrt{\left(\frac{k}{g}\right)}\right] - \arctan\left[u\sqrt{\left(\frac{k}{g}\right)}\right]$$

The time T that particle P takes to reach the highest point of its path is obtained by substituting $u = 0$

So:
$$T = \frac{1}{\sqrt{(kg)}} \arctan\left[U\sqrt{\left(\frac{k}{g}\right)}\right]$$

Using $a = u\dfrac{du}{dy}$ gives:

$$u\frac{du}{dy} = -g - ku^2$$

Integrating:
$$\int \frac{u\,du}{g + ku^2} = -y + A$$

$$\frac{1}{2k} \ln(g + ku^2) = -y + A$$

Substituting $u = U$ when $y = 0$ gives:

$$A = \frac{1}{2k}\ln(g + kU^2)$$

and:

$$y = \frac{1}{2k}\ln\left[\frac{g + kU^2}{g + ku^2}\right]$$

You can find the greatest height H reached by the particle by substituting $u = 0$.

So:

$$H = \frac{1}{2k}\ln\left[1 + \frac{kU^2}{g}\right]$$

3.3 Resisted motion of a vehicle whose engine is working at a constant rate

In Book M2, chapter 3, we considered the motion of vehicles when the resistance to their movement was constant. This work can now be extended to the case where the resistance is a function of the speed of the vehicle.

Consider a vehicle of mass m kg moving along a horizontal road with its engine working at the constant rate h W. Suppose the car is subject to a resistance $f(v)$ N where $v\,\text{m}\,\text{s}^{-1}$ is the speed of the vehicle. The tractive force F N can be obtained from

$$F \times v = h$$

or:

$$F = \frac{h}{v}$$

So the equation of motion of the vehicle is

$$ma = F - f(v)$$
$$= \frac{h}{v} - f(v)$$

(i) Using: $a = \dfrac{dv}{dt}$

gives:

$$m\frac{dv}{dt} = \frac{h}{v} - f(v) \tag{1}$$

(ii) Using: $a = v\dfrac{dv}{dx}$

gives:

$$mv\frac{dv}{dx} = \frac{h}{v} - f(v) \tag{2}$$

If a relationship between v and t is required, use equation (1). If a relationship between v and x is required, use (2). The integration will depend on the function $f(v)$ and this is illustrated in the following example.

Example 3

A car of mass $560\,kg$ moves along a straight road. The magnitude of the resistive force to the motion of the car is $80v\,N$, where $v\,m\,s^{-1}$ is the speed of the car. The engine exerts a constant power of $72\,kW$.

(a) Find the time, in seconds, it takes for the car to accelerate from a speed of $10\,m\,s^{-1}$ to a speed of $20\,m\,s^{-1}$.

(b) Find the distance, in metres, that the car travels in this time.

(a) Since information is required regarding the time, use equation (1) with $f(v) = 80v$, $h = 72\,000$ and $m = 560$. The equation of motion is then

$$560\frac{dv}{dt} = \frac{72\,000}{v} - 80v$$

or:

$$\frac{dv}{dt} = \frac{72\,000 - 80v^2}{560v}$$

$$= \frac{900 - v^2}{7v}$$

Separating the variables and integrating between $v = 10$ and $v = 20$ gives:

$$\int_{10}^{20} \frac{7v}{900 - v^2}\,dv = \int_0^T dt$$

or:

$$-\tfrac{7}{2}\left[\ln\left(900 - v^2\right)\right]_{10}^{20} = T$$

So:

$$T = \tfrac{7}{2}\ln\left[\tfrac{800}{500}\right] = \tfrac{7}{2}\ln\left(\tfrac{8}{5}\right) = 1.65 \text{ seconds}$$

It takes the car 1.65 seconds to accelerate from a speed of $10\,m\,s^{-1}$ to a speed of $20\,m\,s^{-1}$.

(b) Since you are asked to find the distance, use equation (2). The equation of motion is now

$$v\frac{dv}{dx} = \frac{900 - v^2}{7v}$$

Separating the variables and integrating between $v = 10$ and $v = 20$ gives:

$$\int_{10}^{20} \frac{7v^2}{900 - v^2}\,dv = \int_0^X dx = X$$

The left-hand side may be written as:

$$\int_{10}^{20}\left[-7+\frac{6300}{900-v^2}\right]dv$$

$$=\int_{10}^{20}\left[-7+\frac{6300}{60}\left\{\frac{1}{30+v}+\frac{1}{30-v}\right\}\right]dv$$

$$=\left[-7v+\frac{6300}{60}\ln\left(\frac{30+v}{30-v}\right)\right]_{10}^{20}$$

$$=-7(20-10)+105\ln\left(\tfrac{50}{10}\right)-105\ln\left(\tfrac{40}{20}\right)$$

$$=-70+105\ln\left(\tfrac{5}{2}\right)$$

So: $$X=26.2$$

The car travels 26.2 metres in this time.

Exercise 3A

1 A particle P is moving along Ox with an acceleration of $-(2+v)\,\mathrm{m\,s^{-2}}$ at time t seconds when its displacement from O is x metres and its speed is $v\,\mathrm{m\,s^{-1}}$. When $t=0$, $v=12$.
   (a) Calculate the value of t when P comes to rest.
   (b) Calculate v when $t=1.5$.

2 A particle P is moving along Ox with an acceleration of $-(4+3v)\,\mathrm{m\,s^{-2}}$ at time t seconds, when its displacement from O is x metres and its speed is $v\,\mathrm{m\,s^{-1}}$. When $x=0$, $v=4$. Calculate the value of x when P comes to rest.

3 A particle P is moving along Ox with an acceleration of $-(v^2+2)\,\mathrm{m\,s^{-2}}$ at time t seconds when its displacement from O is x metres and its speed is $v\,\mathrm{m\,s^{-1}}$. When $x=0$, $v=10$.
   (a) Find the distance travelled when $v=0$.
   (b) Find the speed of P when $x=1$.

4 A particle P of mass $2\,\mathrm{kg}$ is moving along Ox. At time t seconds the displacement of P from O is x metres and its speed is $v\,\mathrm{m\,s^{-1}}$. The particle moves under the action of a retarding force of magnitude $(12+3v^2)\,\mathrm{N}$. Initially, when $t=0$, $x=0$ and $v=4$.
   (a) Find the time taken by the particle to come to rest.
   (b) Find the distance travelled in this time.

5 The retardation of a train with the power cut off is

$$\left(v^2 + \frac{u^2}{4}\right) \mathrm{m\,s^{-2}}$$

where $v\,\mathrm{m\,s^{-1}}$ is the speed and u is a constant.
Initially $v = u$.
(a) Show that the speed will be halved in a distance
$\frac{1}{2}\ln\left(\frac{5}{2}\right)$ metres in time $\frac{2}{u}\left[\arctan 2 - \frac{\pi}{4}\right]$ seconds.
(b) Show also that the train will come to rest in a further
distance $\frac{1}{2}\ln 2$ metres in additional time $\dfrac{\pi}{2u}$ seconds.

6 A particle P of mass $1\,\mathrm{kg}$ is projected along a rough
horizontal table with speed $u\,\mathrm{m\,s^{-1}}$. The coefficient of friction
between the particle and the surface is μ. The particle also
suffers an air resistance of magnitude $9.8kv^2\,\mathrm{N}$ when $v\,\mathrm{m\,s^{-1}}$
is the speed of the particle and k is a constant. Find the
distance travelled by P before it comes to rest and the time it
takes to come to rest.

7 A ball of mass m is moving vertically under gravity. The air
resistance is assumed to be mkv^2, where v is the speed of the
ball and k is a constant. Initially the ball is projected
downwards with a speed u where $u^2 < \dfrac{g}{k}$.

(a) Find the speed of the ball when it has fallen a distance d.
(b) Compare this with the value you would get for the speed
of the ball at this point if you neglect air resistance.

8 A small stone of mass m moves in air in which the
resistance to its motion varies as the square of the speed.
The magnitude of the terminal speed is V. The stone is
projected vertically upwards with a speed $V\tan\alpha$, where
α is a constant.
(a) Show that the total energy, kinetic plus potential energy,
lost in its ascent is

$$\tfrac{1}{2}mV^2(\tan^2\alpha - 2\ln\sec\alpha)$$

(b) Show also that the stone will return to the point of
projection with a speed $V\sin\alpha$.

9 A ball of mass m kg is projected vertically upwards with a speed of $19.6 \, \text{m s}^{-1}$ in a medium that subjects the ball to a resistance of magnitude $\frac{1}{2}mv$ where $v \, \text{m s}^{-1}$ is the speed of the ball.

(a) Show that t seconds after projection

$$v = 19.6(2e^{-t/2} - 1)$$

(b) Deduce that the ball reaches its maximum height $(\ln 4)$ seconds after projection.

10 A parachutist falls from rest from a balloon. Her parachute opens immediately and the air resistance with the parachute open is proportional to the speed of the parachutist. The terminal speed is $4 \, \text{m s}^{-1}$. Determine the distance the parachutist drops in the first 2 seconds of her fall.

11 A car of mass m is moving along a horizontal road with its engine working at an effective constant rate R. There is a resistance to the motion of the car of magnitude kv^2 where v is the speed of the car. Find the distance travelled by the car as its speed increases from $0.25\left(\dfrac{R}{k}\right)^{\frac{1}{3}}$ to $0.75\left(\dfrac{R}{k}\right)^{\frac{1}{3}}$.

12 A particle P of mass m falls from rest in a medium that produces a resistance of magnitude mkv, where k is a constant, when the speed of the particle is v. Show that when P has reached a speed of V it will have fallen for a time

$$\frac{1}{k}\ln\left(\frac{g}{g - kV}\right)$$

SUMMARY OF KEY POINTS

1 For a particle P travelling in a straight line, which at time t seconds has a displacement x metres from a fixed point O of the line and a velocity of $v \, \text{m s}^{-1}$ the acceleration $a \, \text{m s}^{-2}$ is given by

$$a = \frac{dv}{dt}$$

or:

$$a = v\frac{dv}{dx} = \frac{d}{dx}(\tfrac{1}{2}v^2)$$

2 Displacements must always be measured from a **fixed** point.

Damped and forced harmonic oscillations

4

4.1 Damped harmonic oscillations

In chapter 3 of Book M3 you studied simple harmonic motion. This is motion along a straight line in which the force acting on the particle is directed towards a fixed point of the line and is proportional to the distance of the particle from that fixed point. If the fixed point is O, the distance x m of the particle, P, of mass m kg, from O satisfies the differential equation

$$m\ddot{x} = -m\omega^2 x$$

or:
$$\ddot{x} = -\omega^2 x \qquad (1)$$

at time t seconds.

The general solution of equation (1) (see chapter 3 of Book M3) is

$$x = a\sin(\omega t + \alpha) \qquad (2)$$

where a and α are constants known as the amplitude and phase respectively. The graph of x against t is:

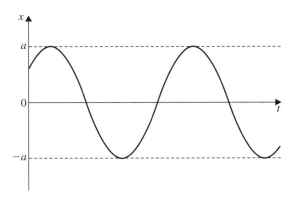

Section 3.4 of Book M3 studied the oscillations of a particle attached to one end of an elastic spring and moving on a smooth horizontal surface. We assumed there that the only horizontal force acting on the particle was the tension in the spring. The extension x of the spring then satisfies equation (1) with

$$\omega^2 = \frac{\lambda}{ml}$$

where l is the natural length and λ the modulus of the spring.

According to the graph on page 71 the particle will perform oscillations of constant amplitude. This is not what is observed in practice. The amplitude of the oscillations decreases in magnitude and they die out fairly quickly. The model must therefore be refined by taking other factors into account. The refinement to be considered here takes air resistance into account.

Consider now the situation where the particle suffers a resistance proportional to its speed $v \, \mathrm{m \, s^{-1}}$ in addition to the tension in the spring. Let the resistance be $mkv \, \mathrm{N}$. The equation of motion is now

$$m \frac{d^2 x}{dt^2} = -m\omega^2 x - mk \frac{dx}{dt}$$

or:

$$\frac{d^2 x}{dt^2} + k \frac{dx}{dt} + \omega^2 x = 0 \qquad (3)$$

This is a second order differential equation with constant coefficients. The solution of differential equations of this type is dealt with in chapter 6 of Book P4. The form of the solution depends on the relative values of k and ω^2. When k is small the motion is said to be **lightly damped**. When k is large the motion is said to be **heavily damped**.

You need to consider three separate cases, depending on whether k^2 is greater than, equal to, or less than $4\omega^2$. These correspond to the auxiliary equation for equation (3) having real distinct, equal, or complex roots.

Substituting $x = e^{\lambda t}$ into equation (3) gives

$$\lambda^2 + k\lambda + \omega^2 = 0 \qquad (4)$$

(the auxiliary equation) so that $e^{\lambda t}$ is a solution provided that

$$\lambda = \frac{-k \pm \sqrt{(k^2 - 4\omega^2)}}{2}$$

(1) $k^2 > 4\omega^2$ (heavy damping)

There are two real distinct roots of equation (4), namely:

$$\lambda_1 = -\frac{k}{2} + \sqrt{\left(\frac{k^2}{4} - \omega^2\right)} < 0$$

and

$$\lambda_2 = -\frac{k}{2} - \sqrt{\left(\frac{k^2}{4} - \omega^2\right)} < 0$$

The general solution is now

$$x = A e^{\lambda_1 t} + B e^{\lambda_2 t} \qquad (5)$$

where A and B are arbitrary constants. Since both λ_1 and λ_2 are negative, both $e^{\lambda_1 t}$ and $e^{\lambda_2 t}$ *decrease to zero* as $t \to \infty$. The graph of x against t depends on the values of the constants A and B which in turn depend on the initial conditions, that is, the values of x and $\dot{x}$ when $t = 0$. The three possibilities are shown here:

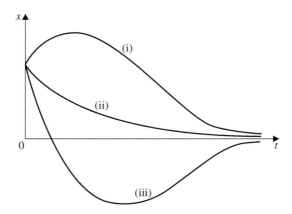

Example 1 will illustrate when each occurs.

(2) $k^2 = 4\omega^2$

The roots of the auxiliary equation are equal.

So: $$\lambda_1 = \lambda_2 = -\frac{k}{2} = -\omega$$

and the general solution is

$$x = (\alpha + \beta t)e^{-\frac{1}{2}kt} = (\alpha + \beta t)e^{-\omega t} \qquad (6)$$

where α and β are arbitrary constants.

This has the same features as the solution in case (1). Again there are three possible cases (i), (ii) and (iii), depending on the initial conditions.

(3) $k^2 < 4\omega^2$ (light damping)

The roots are complex and:

$$\lambda_1 = -\frac{k}{2} + ip$$

$$\lambda_2 = -\frac{k}{2} - ip$$

where $p^2 = \omega^2 - \frac{k^2}{4}$.

The general solution can now be written in the form:

$$x = Ae^{-\frac{kt}{2}}\cos pt + Be^{-\frac{kt}{2}}\sin pt$$

or: $$x = ae^{-\frac{kt}{2}}\cos(pt + \alpha) \qquad (7)$$

where A, B, a and α are arbitrary constants.

The graph of x against t is:

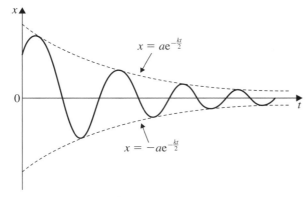

Since x lies between $\pm ae^{-\frac{kt}{2}}$ the graph of x lies between the two curves with equations $x = ae^{-\frac{kt}{2}}$ and $x = -ae^{-\frac{kt}{2}}$, which are also shown in the figure.

Notice the following points.

1. The motion is oscillatory, with the particle passing through the centre O, where $x = 0$, when $\cos(pt + \alpha) = 0$; that is, when $pt + \alpha$ is an odd integral multiple of $\frac{\pi}{2}$, that is $(2n+1)\frac{\pi}{2}$. So these values of t occur at intervals of $\frac{\pi}{p}$.

The period of oscillation is the time to complete one oscillation. Here one complete oscillation is the path travelled by the particle in the interval of time between passing through O in the same direction on successive occasions. The period is then:

$$\frac{2\pi}{p} = \frac{2\pi}{\sqrt{\left(\omega^2 - \frac{k^2}{4}\right)}}$$

Compare this with the period of undamped simple harmonic motion, which is $\frac{2\pi}{\omega}$. The period of oscillation is therefore *lengthened* by the damping.

2. The maximum and minimum values of x occur when $\frac{dx}{dt} = 0$, that is when

$$\tfrac{1}{2}k\cos(pt + \alpha) + p\sin(pt + \alpha) = 0$$

or:

$$\tan(pt + \alpha) = -\frac{k}{2p}$$

The maximum and minimum values therefore occur at intervals of $\frac{\pi}{p}$, since the tangent function is of period π (see chapter 2,

Book P1). It can be shown that their numerical values diminish in geometrical progression with common ratio $e^{-k\pi/2p}$ (see example 2). This case is of practical importance and the term 'damped harmonic motion' is often applied to it.

Example 1

A particle P of mass 1 kg moves in a horizontal straight line under the action of a force directed towards a fixed point O of the line. The force varies as the distance of the particle from O and is equal to $4x$ N when P is at a distance x metres from O. The particle is also subject to a resisting force which is proportional to its speed and which is equal to $5v$ N when the speed of P is $v\,\text{m}\,\text{s}^{-1}$. The particle is at $x = 2$ when $t = 0$. Consider the motion of P in the three cases:

(a) P is projected *away* from O with speed $3\,\text{m}\,\text{s}^{-1}$
(b) P is projected *towards* O with speed $1\,\text{m}\,\text{s}^{-1}$
(c) P is projected *towards* O with speed $9\,\text{m}\,\text{s}^{-1}$.

The equation of motion of the particle is

$$\frac{d^2x}{dt^2} = -4x - 5\frac{dx}{dt}$$

at time t seconds. That is:

$$\frac{d^2x}{dt^2} + 5\frac{dx}{dt} + 4x = 0$$

Substituting $x = Ae^{\lambda t}$ gives:

$$\lambda^2 + 5\lambda + 4 = 0$$

or: $$(\lambda + 4)(\lambda + 1) = 0$$

So: $$\lambda = -1 \text{ or } -4$$

Hence: $$x = Ae^{-4t} + Be^{-t}$$

(a) You can determine the constants A and B by using the initial conditions.

Since $x = 2$ when $t = 0$,

$$2 = A + B \tag{1}$$

As: $$x = Ae^{-4t} + Be^{-t}$$

then: $$\frac{dx}{dt} = -4Ae^{-4t} - Be^{-t}$$

Since: $\dfrac{dx}{dt} = 3$ when $t = 0$,

$$3 = -4A - B \tag{2}$$

Adding equations (1) and (2) gives:

$$5 = -3A$$

so:

$$A = -\tfrac{5}{3} = -1\tfrac{2}{3}$$

From (1):

$$-1\tfrac{2}{3} + B = 2$$

so:

$$B = 2 + 1\tfrac{2}{3} = 3\tfrac{2}{3}$$

The solution for this case is

$$x = -1\tfrac{2}{3}\mathrm{e}^{-4t} + 3\tfrac{2}{3}\mathrm{e}^{-t}$$

In order to sketch x against t note:

(i) $x = 0$ when $-1\tfrac{2}{3}\mathrm{e}^{-4t} + 3\tfrac{2}{3}\mathrm{e}^{-t} = 0$

or:

$$\tfrac{5}{11} = \mathrm{e}^{3t}$$

As $\tfrac{5}{11}$ is less than 1 and e^{3t} is greater than 1 for $t > 0$, this equation has no real positive solution. Hence x **is never zero for $t \geqslant 0$.**

(ii) $\dfrac{\mathrm{d}x}{\mathrm{d}t} = 0$ when

$$+\tfrac{20}{3}\mathrm{e}^{-4t} - \tfrac{11}{3}\mathrm{e}^{-t} = 0$$

or:

$$\tfrac{20}{11} = \mathrm{e}^{3t}$$

As $\tfrac{20}{11}$ is greater than 1 this equation does have a real positive solution:

$$t_0 = \tfrac{1}{3}\ln\left(\tfrac{20}{11}\right)$$

(iii) Both e^{-4t} and e^{-t} tend to zero as $t \to \infty$. The dependence of x on t in this case is shown in the sketch:

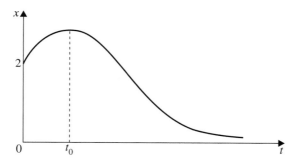

(b) Consider now the general solution:

$$x = A'\mathrm{e}^{-4t} + B'\mathrm{e}^{-t}$$

Since $x = 2$ when $t = 0$, it follows that:

$$2 = A' + B' \tag{3}$$

As P is now directed *towards* O with speed $1\,\mathrm{m\,s^{-1}}$,

$$\frac{\mathrm{d}x}{\mathrm{d}t} = -1 \quad \text{when} \quad t = 0$$

So:
$$-1 = -4A' - B' \qquad (4)$$

Adding equations (3) and (4) gives:

$$1 = -3A'$$

so:
$$A' = -\tfrac{1}{3}$$

From (3):
$$-\tfrac{1}{3} + B' = 2$$

so:
$$B' = 2 + \tfrac{1}{3} = 2\tfrac{1}{3}$$

and the solution for this case is

$$x = -\tfrac{1}{3}\mathrm{e}^{-4t} + 2\tfrac{1}{3}\mathrm{e}^{-t}$$

In order to sketch x against t note:

(i) $x = 0$ when $-\tfrac{1}{3}\mathrm{e}^{-4t} + 2\tfrac{1}{3}\mathrm{e}^{-t} = 0$

or:
$$\tfrac{1}{7} = \mathrm{e}^{3t}$$

As before, since $\tfrac{1}{7}$ is less than 1 this equation has no real positive solution. Hence **x never vanishes for $t \geqslant 0$**.

(ii) $\dfrac{\mathrm{d}x}{\mathrm{d}t} = 0$ when $\tfrac{4}{3}\mathrm{e}^{-4t} - \tfrac{7}{3}\mathrm{e}^{-t} = 0$

or:
$$\tfrac{4}{7} = \mathrm{e}^{3t}$$

As $\tfrac{4}{7}$ is less than 1 this equation has no real positive solution. Hence $\dfrac{\mathbf{d}x}{\mathbf{d}t}$ **never vanishes for $t \geqslant 0$**. The sketch of x against t in this case is:

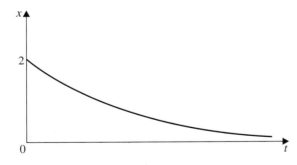

(c)   Consider now the general solution:

$$x = A''\mathrm{e}^{-4t} + B''\mathrm{e}^{-t}$$

Since $x = 2$ when $t = 0$, it follows that:

$$2 = A'' + B'' \qquad (5)$$

As P is now directed *towards* O with speed $9 \, \text{m} \, \text{s}^{-1}$,

$$\frac{\text{d}x}{\text{d}t} = -9 \quad \text{when} \quad t = 0$$

so:
$$-9 = -4A'' - B'' \tag{6}$$

Adding equations (5) and (6) gives:

$$-7 = -3A''$$

so:
$$A'' = \tfrac{7}{3}$$

From (5):
$$\tfrac{7}{3} + B'' = 2$$

so:
$$B'' = -\tfrac{1}{3}$$

and the solution for this case is

$$x = \tfrac{7}{3}\text{e}^{-4t} - \tfrac{1}{3}\text{e}^{-t}$$

In order to sketch x against t note:

(i) $x = 0$ when $\tfrac{7}{3}\text{e}^{-4t} - \tfrac{1}{3}\text{e}^{-t} = 0$.

So:
$$7 = \text{e}^{3t}$$

or:
$$t_1 = \tfrac{1}{3}\ln 7$$

(ii) $\dfrac{\text{d}x}{\text{d}t} = 0$ when $-\tfrac{28}{3}\text{e}^{-4t} + \tfrac{1}{3}\text{e}^{-t} = 0$.

So:
$$28 = \text{e}^{3t}$$

or:
$$t_2 = \tfrac{1}{3}\ln 28 > t_1$$

The sketch of x against t in this case is:

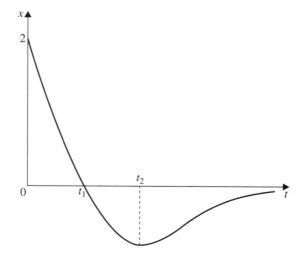

Example 2

A particle P is moving along a horizontal straight line so that at time t seconds its distance x m from a fixed point O of the line satisfies the differential equation:

$$4\frac{d^2x}{dt^2} + 12\frac{dx}{dt} + 13x = 0$$

(a) Describe a physical situation that could give rise to this equation.

(b) Given that $x = 4$ and $\dfrac{dx}{dt} = 6$ when $t = 0$, obtain the solution of this differential equation.

(c) Hence show that the motion is oscillatory with constant period.

(d) Show also that the amplitudes of successive oscillations decrease in geometric progression.

(a) The given equation can be written:

$$4\frac{d^2x}{dt^2} = -13x - 12\frac{dx}{dt}$$

or:

$$\frac{d^2x}{dt^2} = -\tfrac{13}{4}x - 3\frac{dx}{dt}$$

The left-hand side is just the acceleration of the particle P, which can be taken to be of unit mass. Then the term $-\tfrac{13}{4}x$ is a force whose magnitude is proportional to the distance x metres from a fixed point O of the line of motion. It is directed towards O. The term $-3\dfrac{dx}{dt}$ is a force whose magnitude is proportional to the speed but with direction opposite to the direction of motion. It therefore may be taken as a force resisting the motion.

(b) Substituting $x = Ae^{\lambda t}$ into the differential equation gives the auxiliary equation:

$$4\lambda^2 + 12\lambda + 13 = 0$$

The roots of this equation are

$$\lambda = \frac{-12 \pm \sqrt{[(12)^2 - 4 \times 4 \times 13]}}{2 \times 4}$$

$$= -\tfrac{3}{2} \pm i$$

So the general solution is

$$x = Ae^{-\frac{3}{2}t}\cos t + Be^{-\frac{3}{2}t}\sin t \tag{1}$$

Substituting $x = 4$ and $t = 0$ gives:

$$4 = A + 0$$

so:

$$A = 4$$

From (1):

$$\frac{dx}{dt} = A[-\tfrac{3}{2}e^{-\frac{3}{2}t}\cos t + e^{-\frac{3}{2}t}(-\sin t)] + B[-\tfrac{3}{2}e^{-\frac{3}{2}t}\sin t + e^{-\frac{3}{2}t}\cos t]$$

Substituting $\dfrac{dx}{dt} = 6$ and $t = 0$ gives:

$$-\tfrac{3}{2}A + B = 6$$

so:
$$B = \tfrac{3}{2}A + 6 = 12$$

and hence:
$$x = 4e^{-\frac{3}{2}t}\cos t + 12e^{-\frac{3}{2}t}\sin t$$

This may be written as

$$x = e^{-\frac{3}{2}t}[4\cos t + 12\sin t]$$

If:
$$4\cos t + 12\sin t = r\cos(t + \alpha)$$
$$= r\cos t\cos\alpha - r\sin t\sin\alpha$$

then:
$$r\cos\alpha = 4 \quad\text{and}\quad r\sin\alpha = -12$$
$$r^2\cos^2\alpha = 16 \quad\text{and}\quad r^2\sin^2\alpha = 144$$

$$r^2(\cos^2\alpha + \sin^2\alpha) = r^2 = 16 + 144 = 160$$

So:
$$r = \sqrt{160} = 4\sqrt{10}$$

$$\tan\alpha = \frac{r\sin\alpha}{r\cos\alpha} = \frac{-12}{4} = -3$$

giving:
$$x = 4\sqrt{(10)}e^{-\frac{3t}{2}}\cos(t + \alpha)$$

where $\tan\alpha = -3$.

(c) This solution involves $\cos(t + \alpha)$ which is an oscillating function. It takes the value zero when

$$t + \alpha = (2n + 1)\frac{\pi}{2}$$

where n is an integer. The values of t that satisfy this occur at intervals of π and so the motion is periodic.

(d) $\dfrac{dx}{dt} = 0$ when

$$-\tfrac{3}{2}e^{-\frac{3t}{2}}\cos(t + \alpha) - e^{-\frac{3t}{2}}\sin(t + \alpha) = 0$$

or:
$$\tan(t + \alpha) = -\tfrac{3}{2}$$

$\Rightarrow$
$$t + \alpha = \arctan\left(-\tfrac{3}{2}\right) + N\pi$$

where N is an integer.

$$t = \arctan\left(-\tfrac{3}{2}\right) - \arctan(-3) + N\pi$$
$$= \beta + N\pi \text{ (say)}$$

Therefore x has stationary values, which are alternately maxima and minima, for values of t given by the equation above. The maxima are given by the even values of N, giving

$$t = \beta + 2n\pi$$

and:
$$\cos^2(t + \alpha) = \frac{1}{\sec^2(t + \alpha)}$$

$$= \frac{1}{1 + \tan^2(t + \alpha)}$$

$$= \frac{1}{1 + \frac{9}{4}} = \frac{4}{13}$$

So:
$$\cos(t + \alpha) = \frac{2}{\sqrt{13}}$$

Then successive values of the amplitude are:

$$a_1 = 4\sqrt{10} \times \frac{2}{\sqrt{13}} e^{-\frac{3}{2}(\beta + 2\pi)} = (8\sqrt{\tfrac{10}{13}}) e^{-\frac{3\beta}{2} - 3\pi}$$

$$a_2 = 4\sqrt{10} \times \frac{2}{\sqrt{13}} e^{-\frac{3}{2}(\beta + 4\pi)} = (8\sqrt{\tfrac{10}{13}}) e^{-\frac{3\beta}{2} - 6\pi}$$

$$a_3 = 4\sqrt{10} \times \frac{2}{\sqrt{13}} e^{-\frac{3}{2}(\beta + 6\pi)} = (8\sqrt{\tfrac{10}{13}}) e^{-\frac{3\beta}{2} - 9\pi}$$

and so on.

From this

$$\frac{a_{n+1}}{a_n} = e^{-3\pi}$$

and so the amplitudes decrease in geometric progression.

Example 3

A particle P of mass $2\,\text{kg}$ moves along a horizontal straight line under the action of a force directed towards a fixed point O of the line. The force has magnitude $8x\,\text{N}$ when P is at a distance x metres from O. The particle also suffers air resistance which is proportional to its speed and which has magnitude $8v\,\text{N}$ when the speed of P is $v\,\text{m\,s}^{-1}$. Given that when $t = 0$ the particle is at rest and $x = a$, find the greatest speed of the particle in the ensuing motion.

The equation of motion of P is

$$2\frac{\mathrm{d}^2 x}{\mathrm{d}t^2} = -8x - 8\frac{\mathrm{d}x}{\mathrm{d}t}$$

or:
$$\frac{\mathrm{d}^2 x}{\mathrm{d}t^2} + 4\frac{\mathrm{d}x}{\mathrm{d}t} + 4x = 0$$

at time t seconds.

Substituting $x = Ae^{\lambda t}$:

$$\lambda^2 + 4\lambda + 4 = 0$$

or:
$$(\lambda + 2)^2 = 0$$

So:
$$\lambda = -2 \quad \text{twice}$$

Hence:
$$x = e^{-2t}(A + Bt) \tag{1}$$

Substituting $x = a$ and $t = 0$ gives:

$$a = A$$

From (1):
$$\frac{\mathrm{d}x}{\mathrm{d}t} = -2e^{-2t}A + Be^{-2t} - 2Bte^{-2t}$$

Substituting $\dot{x} = 0$ and $t = 0$ gives:

$$0 = -2A + B$$

So:
$$B = 2A = 2a$$

and:
$$x = ae^{-2t}(1 + 2t) \tag{2}$$

The maximum speed of P occurs when $\ddot{x} = 0$.

From (2):
$$\dot{x} = -2ae^{-2t}(1 + 2t) + 2ae^{-2t}$$
$$= -4ate^{-2t}$$

and
$$\ddot{x} = -4ae^{-2t} + 8ate^{-2t}$$

When $\ddot{x} = 0$:
$$-4a + 8at = 0$$

So:
$$t = \tfrac{1}{2}$$

When $t = \tfrac{1}{2}$:
$$\dot{x} = -4a \times \tfrac{1}{2} \times e^{-2 \times \frac{1}{2}} = -2ae^{-1}$$

and so the maximum speed of P is $2ae^{-1}\,\mathrm{m\,s^{-1}}$.

Example 4

The elastic strings AB and BC are each of natural length a and modulus of elasticity $\frac{3}{2}mg$ and are joined at B. A particle of mass m is attached to the strings at B. The ends A and C of the strings are fixed to points, on a smooth horizontal plane, which are a distance $4a$ apart. The particle is set in motion and moves in a straight line along the line AC. Its displacement x is measured from the mid-point of the line AC. The motion takes place in a medium which offers a resistance of magnitude mnv, where v is the speed of the particle and $n = \sqrt{\left(\dfrac{g}{a}\right)}$.

(a) Show that $\dfrac{\mathrm{d}^2 x}{\mathrm{d}t^2} + n\dfrac{\mathrm{d}x}{\mathrm{d}t} + 3n^2 x = 0$.

Given that when $t = 0$, $x = a$ and $\dfrac{\mathrm{d}x}{\mathrm{d}t} = \tfrac{1}{2}na$,

(b) find an expression for x in terms of a, n and t.

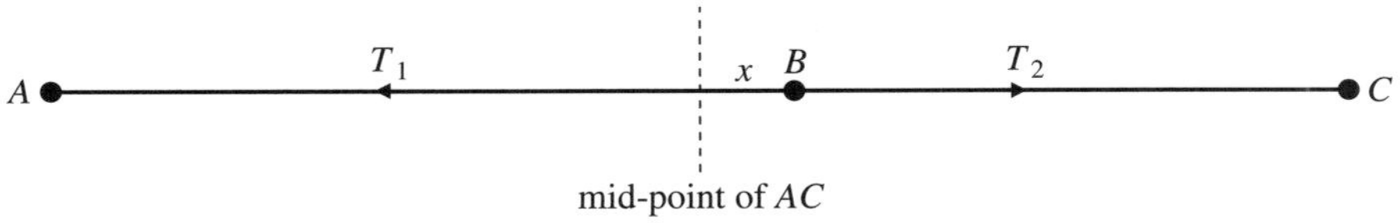

mid-point of AC

The diagram shows the given information with the particle B displaced a distance x from the mid-point of AC. Then since $AC = 4a$ it follows that $AB = 2a + x$ and $BC = 2a - x$. Let T_1 and T_2 be the tensions in the strings AB and BC.

Using Hooke's law $T = \dfrac{\lambda}{l}$ (extension) (see Book M3, chapter 2)

$$T_1 = \frac{3mg}{2}\frac{(a + x)}{a}$$

since the extension of AB is $(a + x)$ and its initial length is a.

Similarly
$$T_2 = \frac{3mg}{2}\frac{(a - x)}{a}$$

The equation of motion of B is then

$$m\ddot{x} = T_2 - T_1 - mnv$$
$$= \frac{3mg}{2}\frac{(a + x)}{a} - \frac{3mg}{2}\frac{(a - x)}{a} - mn\dot{x}$$
$$= \frac{3mg}{a}x - mn\dot{x}$$

So:
$$\ddot{x} + n\dot{x} + \frac{3g}{a}x = 0$$

or:
$$\ddot{x} + n\dot{x} + 3n^2 x = 0 \quad \text{as} \quad n^2 = \frac{g}{a}.$$

(b) The differential equation obtained in part (a) has constant coefficients. The corresponding auxiliary equation is

$$\lambda^2 + n\lambda + 3n^2 = 0$$

which has solutions

$$\lambda = \frac{-n \pm \sqrt{(n^2 - 12n^2)}}{2}$$
$$= -\frac{n}{2} \pm \frac{in}{2}\sqrt{11}.$$

So the general solution is

$$x = e^{-\frac{nt}{2}}\left(A\cos\tfrac{\sqrt{11}}{2}nt + B\sin\tfrac{\sqrt{11}}{2}nt\right)$$

(See Book P4, chapter 6.)

The condition $x = a$ when $t = 0$ gives

$$a = A$$

To use the other condition we need

$$\dot{x} = -\tfrac{n}{2}x + e^{-\frac{nt}{2}}\left(-an\tfrac{\sqrt{11}}{2}\sin\tfrac{\sqrt{11}}{2}nt + \tfrac{\sqrt{11}}{2}nB\cos\tfrac{\sqrt{11}}{2}nt\right)$$

Using: $\dot{x} = \tfrac{1}{2}na$ and $x = a$ when $t = 0$ gives:

$$\tfrac{1}{2}na = -\tfrac{n}{2}a + \tfrac{\sqrt{11}}{2}nB$$

So $B = \dfrac{2a}{\sqrt{11}}$.

Hence:

$$x = e^{-\frac{nt}{2}}\left(a\cos\tfrac{\sqrt{11}}{2}nt + \tfrac{2a}{\sqrt{11}}\sin\tfrac{\sqrt{11}}{2}nt\right).$$

Example 5

A particle P of mass m is attached to one end of an elastic string of modulus mg and natural length l. The other end of the string is attached to a fixed point O. The particle is held below O so that the string is vertical and of length $3l$ and is then released. The subsequent motion takes place in a medium which offers a resistance of magnitude $\left(\dfrac{3g}{l}\right)^{\frac{1}{2}}mv$, where v is the speed of the particle.

(a) Show that the extension x of the string at time t satisfies the differential equation

$$l\frac{d^2x}{dt^2} + (3gl)^{\frac{1}{2}}\frac{dx}{dt} + gx = gl.$$

(b) Prove that the string does not become slack and that the particle is next at rest when its depth below O is $l(2 - e^{-\sqrt{3}\pi})$. [E]

(a) The diagram shows the situation when the extension of the string is x. When the particle P is in equilibrium the tension in the string T_E is equal to mg

$$T_E = mg$$

If e is the extension then by Hooke's law $T_E = \dfrac{mge}{l}$.

Combining these two equations gives $e = l$.

In the general position shown in the diagram the equation of motion is

$$m\ddot{x} = mg - T + mv\left(\frac{3g}{l}\right)^{\frac{1}{2}}$$

and now $T = \dfrac{mgx}{l}$ and $v = -\dot{x}$

so:
$$m\ddot{x} = mg - mg\frac{x}{l} - m\dot{x}\left(\frac{3g}{l}\right)^{\frac{1}{2}}$$

or:
$$l\ddot{x} + (3gl)^{\frac{1}{2}}\dot{x} + gx = gl.$$

(b) A particular solution of this differential equation is $x_{\mathrm{p}} = l$. To find the complementary function we use the auxiliary equation
$$l\lambda^2 + (3gl)^{\frac{1}{2}}\lambda + g = 0$$

By defining $n^2 = \dfrac{g}{4l}$ this may be written
$$\lambda^2 + 2\sqrt{3}n\lambda + 4n^2 = 0$$

The roots of this equation are:
$$\lambda = \frac{-2\sqrt{3}n \pm \sqrt{12n^2 - 16n^2}}{2} = -\sqrt{3}n \pm in$$

The complementary function is then:
$$x_{\mathrm{c}} = e^{-\sqrt{3}nt}\alpha\cos(nt + \varepsilon)$$

and the general solution of the differential equation found in (a) is
$$x = l + e^{-\sqrt{3}nt}\alpha\cos(nt + \varepsilon)$$

where α and ε are constants.

Initially P is a distance $3l$ below O so that $x = 2l$.

So:
$$2l = l + \alpha\cos\varepsilon$$

or:
$$\alpha\cos\varepsilon = l$$

Since $\dot{x} = 0$ when $t = 0$ we need to obtain $\dot{x}$
$$\dot{x} = -\sqrt{3}ne^{-\sqrt{3}nt}\alpha\cos(nt + \varepsilon) - e^{-\sqrt{3}nt}n\alpha\sin(nt + \varepsilon)$$

Then:
$$0 = -\sqrt{3}n\alpha\cos\varepsilon - n\alpha\sin\varepsilon$$

Using $\alpha\cos\varepsilon = l$ obtained above, gives:
$$-\sqrt{3}l - \alpha\sin\varepsilon = 0$$

so:
$$\alpha\sin\varepsilon = -\sqrt{3}l$$

and:
$$\tan\varepsilon = -\sqrt{3} \quad \text{and} \quad \alpha^2 = l^2 + 3l^2$$

So:
$$\varepsilon = -\tfrac{\pi}{3} \quad \text{and} \quad \alpha = 2l$$

So:
$$x = l + 2le^{-\sqrt{3}nt}\cos\left(nt - \tfrac{\pi}{3}\right)$$

and the motion of P is damped harmonic motion.

To find the greatest height reached in the subsequent oscillation we need to find the time when $\dot{x} = 0$. This occurs when

$$0 = 2l(-\sqrt{3}n)e^{-\sqrt{3}nt}\cos\left(nt - \tfrac{\pi}{3}\right) - e^{-\sqrt{3}nt}2ln\sin\left(nt - \tfrac{\pi}{3}\right)$$

So:

$$\tan\left(nt - \tfrac{\pi}{3}\right) = -\sqrt{3}$$

and $nt - \tfrac{\pi}{3} = -\tfrac{\pi}{3}, \tfrac{2\pi}{3}, \ldots$

so: $nt \quad = 0, \pi, \ldots$

When $t = \dfrac{\pi}{n}$ then $x = l + 2le^{-\sqrt{3}\pi}\cos\dfrac{2\pi}{3}$

$$= l - le^{-\sqrt{3}\pi}$$

The distance of P below O is then

$$l + (l - le^{-\sqrt{3}\pi}) = l(2 - e^{-\sqrt{3}\pi})$$

Since this is greater than l the string will never become slack.

4.2 Forced harmonic oscillations

In the previous section you saw that, under certain circumstances, a particle subject to two forces – one of which was a restoring force proportional to the particle's displacement and the other of which was a resisting force proportional to its speed – would perform oscillations with a certain period. But what happens if you try to make such a system move in a different way, for example trying to force it to oscillate with a frequency (period) other than its natural one?

Suppose, as before, that the particle is of mass m kg, the restoring force is $m\omega^2 x$ N, where x m is the distance from a fixed point O in the line of motion, and the resistance is mkv N, where v is the speed of the particle. Suppose also that there is an additional applied force of magnitude $mf(t)$ N which varies with time. The equation of motion is now

$$m\ddot{x} = -m\omega^2 x - mk\frac{dx}{dt} + mf(t)$$

or:

$$\ddot{x} + k\dot{x} + \omega^2 x = f(t) \tag{1}$$

The solution of this equation depends on the particular form of the function $f(t)$. However, it can be shown that the solution can always be written as the sum of the general solution of

$$\ddot{x} + k\dot{x} + \omega^2 x = 0$$

called the **complementary equation**, and an additional term. (See Book P4 chapter 6.)

Suppose one solution of (1) (called a **particular integral**) is

$$x = u$$

Then: $$\ddot{u} + k\dot{u} + \omega^2 u = f(t)$$

and so equation (1) can be written

$$\ddot{x} + k\dot{x} + \omega^2 x = \ddot{u} + k\dot{u} + \omega^2 u$$

or: $$(\ddot{x} - \ddot{u}) + k(\dot{x} - \dot{u}) + \omega^2(x - u) = 0$$

That is, $y = x - u$ satisfies

$$\ddot{y} + k\dot{y} + \omega^2 y = 0$$

the complementary equation.

It follows that the general solution of (1) is

$$x = u + y$$

where u is a particular integral of (1) and y is the general solution of the complementary equation, usually called the **complementary function**.

So:

■ $x =$ (**particular integral**) $+$ (**complementary function**)

The form of the complementary function for the various cases that can arise is given in the previous section. Methods of finding particular integrals are discussed in chapter 6 of Book P4. The following examples indicate the method of solution in the cases:

$$f(t) = A \cos nt + B \sin nt$$
$$f(t) = \alpha + \beta t$$
$$f(t) = k e^{at}$$

Example 6

A particle P of mass m moves on a fixed straight line. At time t, P is a distance x from a fixed point O of the line. The particle is acted on by three forces:

(i) a force $m\omega^2 x$ directed towards O
(ii) a resistance to the motion of magnitude $2m\omega v$, where v is the speed of P
(iii) a force $F = ma \cos \omega t$ acting in the direction of increasing x.

Given that at $t = 0$ the particle is at rest at O, find x as a function of t. Discuss the motion of P.

The equation of motion of P is

$$m\ddot{x} = -m\omega^2 x - 2m\omega\dot{x} + ma \cos \omega t$$

or: $$\ddot{x} + 2\omega\dot{x} + \omega^2 x = a \cos \omega t \qquad (1)$$

The complementary function is the general solution of the complementary equation:

$$\ddot{x} + 2\omega\dot{x} + \omega^2 x = 0$$

Since the auxiliary equation is

$$\lambda^2 + 2\omega\lambda + \omega^2 = 0$$

or:

$$(\lambda + \omega)^2 = 0$$

the complementary function is

$$e^{-\omega t}(A + Bt)$$

where A and B are arbitrary constants.

To find the particular integral try

$$x_p = p\cos\omega t + q\sin\omega t \qquad (2)$$

where p and q are to be found so that x_p satisfies equation (1).

From (2): $\qquad \dot{x}_p = -p\omega\sin\omega t + q\omega\cos\omega t$

and: $\qquad \ddot{x}_p = -p\omega^2\cos\omega t - q\omega^2\sin\omega t$

Substituting x_p, $\dot{x}_p$ and $\ddot{x}_p$ into (1) gives:

$$(-p\omega^2\cos\omega t - q\omega^2\sin\omega t) + 2\omega(-p\omega\sin\omega t + q\omega\cos\omega t)$$

$$+ \omega^2(p\cos\omega t + q\sin\omega t) = a\cos\omega t$$

Since this equation must hold for all values of t, you can equate the coefficients of $\sin\omega t$ and $\cos\omega t$ on the two sides of the equation.

$\cos\omega t$: $\qquad -p\omega^2 + 2\omega^2 q + \omega^2 p = a \qquad (3a)$

$\sin\omega t$: $\qquad -q\omega^2 - 2\omega^2 p + \omega^2 q = 0 \qquad (3b)$

From (3b): $\qquad p = 0$

From (3a): $\qquad q = \dfrac{a}{2\omega^2}$

Hence: $\qquad x_p = \dfrac{a}{2\omega^2}\sin\omega t$

and so the general solution is

$$x = e^{-\omega t}(A + Bt) + \frac{a}{2\omega^2}\sin\omega t \qquad (4)$$

Now you can find the values of A and B using the initial conditions $x = 0$ and $\dot{x} = 0$ when $t = 0$.

Since $x = 0$ when $t = 0$

$$0 = A$$

From (4), with $A = 0$:

$$\dot{x} = Be^{-\omega t} - \omega Bte^{-\omega t} + \frac{a\omega}{2\omega^2}\cos\omega t$$

Since $\dot{x} = 0$ when $t = 0$:

$$0 = B + \frac{a}{2\omega}$$

so:

$$B = -\frac{a}{2\omega}$$

and

$$x = -\frac{at}{2\omega}e^{-\omega t} + \frac{a}{2\omega^2}\sin\omega t$$

The part of the motion that is represented by $\dfrac{a}{2\omega^2}\sin\omega t$ is an oscillatory motion of constant amplitude $\dfrac{a}{2\omega^2}$ and period $\dfrac{2\pi}{\omega}$, which is the period of the forcing term F. It is therefore called the **forced oscillation**. The part $-\dfrac{at}{2\omega}e^{-\omega t}$ tends to zero as $t \to \infty$, as shown earlier. It is called the **transient part** of the motion.

Example 7

A particle P moves along the x-axis so that at time t its displacement from the origin O satisfies the differential equation

$$\frac{d^2x}{dt^2} + 4\frac{dx}{dt} + 13x = 40\cos 3t$$

(a) Describe a physical situation for which this could be the equation of motion.

(b) Given that $x = 2$ and $\dfrac{dx}{dt} = 13$ when $t = 0$ find x as a function of t.

(c) Show that when t is large the motion approximates to simple harmonic motion about O with period $\dfrac{2\pi}{3}$ and amplitude $\sqrt{10}$.

(a) If the given equation is rearranged in the form

$$\ddot{x} = -13x - 4\frac{dx}{dt} + 40\cos 3t$$

then the left-hand side is the acceleration of a particle of unit mass. The first term on the right-hand side represents a restoring force with magnitude proportional to the displacement. The second term represents a resistance to the motion of P proportional to the speed of P. The final term represents an applied periodic force of period $\dfrac{2\pi}{3}$.

(b) The complementary equation is:

$$\ddot{x} + 4\dot{x} + 13x = 0$$

with auxiliary equation

$$\lambda^2 + 4\lambda + 13 = 0$$

The roots of the auxiliary equation are

$$\lambda = \frac{-4 \pm \sqrt{(16 - 4 \times 13)}}{2} = -2 \pm 3i$$

Hence the complementary function is

$$x_c = e^{-2t}(A\cos 3t + B\sin 3t)$$

where A and B are arbitrary constants.

To find a particular solution try

$$x_p = p\cos 3t + q\sin 3t$$

where p and q are to be found so that x_p satisfies the differential equation.

Differentiating x_p with respect to t gives:

$$\dot{x}_p = -3p\sin 3t + 3q\cos 3t$$

and: $\qquad\qquad\qquad \ddot{x}_p = -9p\cos 3t - 9q\sin 3t$

Substituting x_p, $\dot{x}_p$ and $\ddot{x}_p$ into the differential equation gives:

$$(-9p\cos 3t - 9q\sin 3t) + 4(-3p\sin 3t + 3q\cos 3t)$$
$$+ 13(p\cos 3t + q\sin 3t) = 40\cos 3t$$

Equating coefficients:

$\cos 3t$: $\qquad\qquad -9p + 12q + 13p = 40 \qquad\qquad$ (1)

$\sin 3t$: $\qquad\qquad -9q - 12p + 13q = 0 \qquad\qquad$ (2)

From (2): $\qquad\qquad -12p + 4q = 0$

or $\qquad\qquad\qquad\qquad q = 3p$

From (1): $\qquad\qquad 12q + 4p = 40$

and substituting for q:

$$12 \times 3p + 4p = 40$$

So: $\qquad\qquad\qquad p = 1 \quad \text{and} \quad q = 3$

and the particular integral is

$$x_p = \cos 3t + 3\sin 3t$$

The general solution is therefore

$$x = x_c + x_p$$
$$= e^{-2t}(A\cos 3t + B\sin 3t) + (\cos 3t + 3\sin 3t)$$

To find A and B you must use the initial conditions.

Since $x = 2$ when $t = 0$,

$$2 = A + 1$$

so:

$$A = 1$$

From the above,

$$\dot{x} = -2e^{-2t}(A \cos 3t + B \sin 3t) + e^{-2t}(-3A \sin 3t + 3B \cos 3t)$$

$$- 3 \sin 3t + 9 \cos 3t$$

Since $\dot{x} = 13$ when $t = 0$,

$$-2A + 3B + 9 = 13$$

Using $A = 1$ gives:

$$3B = 13 - 9 + 2$$

So:

$$B = 2$$

and:

$$x = e^{-2t}(\cos 3t + 2 \sin 3t) + (\cos 3t + 3 \sin 3t)$$

(c) When $t \to \infty$ the term $e^{-2t}(\cos 3t + 2 \sin 3t)$, which arises from the complementary equation and is therefore sometimes called the **free oscillation**, tends to zero by virtue of the factor e^{-2t}. So for large values of t the motion is given by

$$x_p = \cos 3t + 3 \sin 3t$$

You can write this in the form

$$x_p = r \cos(3t + \alpha)$$

$$= r \cos 3t \cos \alpha - r \sin 3t \sin \alpha$$

So:

$$r \cos \alpha = 1 \quad \text{and} \quad r \sin \alpha = -3$$

$$r^2 = 10 \quad \text{and} \quad \tan \alpha = -3$$

So:

$$x_p = \sqrt{10} \cos(3t + \alpha)$$

This is simple harmonic motion about O of period $\dfrac{2\pi}{3}$ and amplitude $\sqrt{10}$.

Example 8

A particle P of mass m is attached to one end of a light elastic string of natural length l and modulus of elasticity $6mn^2l$, where n is a constant, lying on a horizontal table. The other end of the string is attached to a fixed point O of the horizontal table and P is initially at rest on the table with $OP = l$. A time-dependent force is now applied to P in the direction OP. The magnitude of the force is mn^2le^{-nt} where t is measured from the time the force was initially applied. The motion of the particle is opposed by a resistance of magnitude $5mnv$ where v is the speed of P. Show that when the extension of the string is x,

$$\frac{d^2x}{dt^2} + 5n\frac{dx}{dt} + 6n^2x = n^2le^{-nt}$$

Show that the greatest extension of the string occurs when $t = \frac{1}{n}\ln 3$ and find this greatest extension.

Using Hooke's law (Book M3, section 2.1) the tension in the string when the extension is x is $6mn^2l\left(\frac{x}{l}\right)$. Therefore for the given situation the equation of motion is

$$m\frac{d^2x}{dt^2} = -6mn^2x - 5mn\frac{dx}{dt} + mn^2le^{-nt}$$

or:

$$\frac{d^2x}{dt^2} + 5n\frac{dx}{dt} + 6n^2x = n^2le^{-nt} \qquad (1)$$

The complementary equation is

$$\ddot{x} + 5n\dot{x} + 6n^2x = 0$$

with auxiliary equation

$$\lambda^2 + 5n\lambda + 6n^2 = 0$$

This equation factorises to give:

$$(\lambda + 3n)(\lambda + 2n) = 0$$

and so has roots $\lambda = -3n$ and $\lambda = -2n$.

Hence the complementary function is

$$x_c = Ae^{-3nt} + Be^{-2nt}$$

where A and B are arbitrary constants.

To find a particular solution try

$$x_p = Ce^{-nt}$$

where C is a constant to be found.

Then:

$$\dot{x}_p = -Cne^{-nt}$$

and:

$$\ddot{x}_p = Cn^2e^{-nt}$$

Substituting into equation (1) gives:

$$C(n^2 - 5n^2 + 6n^2)e^{-nt} = n^2le^{-nt}$$

Since e^{-nt} is not zero,

$$2n^2C = n^2l$$

and

$$C = \frac{l}{2}$$

so that

$$x_p = \frac{l}{2}e^{-nt}$$

Hence the general solution of (1) is

$$x = Ae^{-3nt} + Be^{-2nt} + \tfrac{1}{2}le^{-nt} \qquad (2)$$

You can find the constants A and B by using the initial conditions $x = 0$ and $\dot{x} = 0$ when $t = 0$.

Since $x = 0$ when $t = 0$,

$$0 = A + B + \tfrac{1}{2}l \qquad (3)$$

From (2):
$$\dot{x} = -3nAe^{-3nt} - 2nBe^{-2nt} - \tfrac{1}{2}lne^{-nt}$$

Substituting $\dot{x} = 0$ when $t = 0$ gives:

$$0 = -3nA - 2nB - \tfrac{1}{2}ln \qquad (4)$$

Solving equations (3) and (4) simultaneously gives:

$$A = \tfrac{1}{2}l \quad \text{and} \quad B = -l$$

so that:

$$x = \tfrac{1}{2}le^{-3nt} - le^{-2nt} + \tfrac{1}{2}le^{-nt} \qquad (5)$$

The greatest extension of the string occurs when $\dot{x} = 0$. That is:

$$-\tfrac{3}{2}nle^{-3nt} + 2nle^{-2nt} - \tfrac{1}{2}lne^{-nt} = 0$$

or:

$$\left(\frac{-nl}{2}\right)e^{-3nt}[3 - 4e^{nt} + e^{2nt}] = 0$$

so that:

$$e^{2nt} - 4e^{nt} + 3 = 0$$

This factorises to give:

$$(e^{nt} - 3)(e^{nt} - 1) = 0$$

so that:

$$e^{nt} = 1 \quad \text{or} \quad 3$$

that is:

$$t = 0 \quad \text{or} \quad \frac{1}{n}\ln 3$$

When $t = 0$, $x = 0$ and when $t = \dfrac{1}{n}\ln 3$ the extension of the string is a maximum.

When $t = \dfrac{1}{n}\ln 3$, $e^{-nt} = \frac{1}{3}$ and so:

$$x = \tfrac{1}{2}l(\tfrac{1}{3})^3 - l(\tfrac{1}{3})^2 + \tfrac{1}{2}l(\tfrac{1}{3})$$
$$= l[\tfrac{1}{54} - \tfrac{1}{9} + \tfrac{1}{6}]$$
$$= \dfrac{l}{54}[1 - 6 + 9]$$
$$= \tfrac{2}{27}l$$

So the greatest extension is $\tfrac{2}{27}l$.

Example 9

A light string AB having natural length l and modulus of elasticity $3lmn^2$ lies straight and at its natural length at rest on a horizontal table. A particle of mass m is attached to the end A. The end B is then moved in a straight line in the direction AB with constant acceleration f. The resulting motion of the particle is resisted by a force of magnitude $4mnv$ where v is the speed of the particle. If x is the extension of the string at time t show that

$$\frac{d^2x}{dt^2} + 4n\frac{dx}{dt} + 3n^2x = 4nft + f$$

Find x as a function of t.

Suppose the situation at time t, relative to the initial position, is as shown:

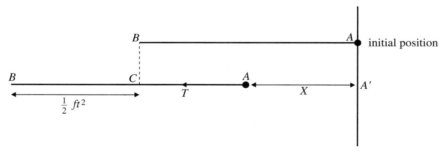

Since B moves in the direction AB with constant acceleration f, then:

$$CB = \tfrac{1}{2}ft^2$$

If A has moved a distance X from its original position A' then the length of the string is

$$l + \tfrac{1}{2}ft^2 - X$$

and so the extension x of the string is

$$x = (l + \tfrac{1}{2}ft^2 - X) - l = \tfrac{1}{2}ft^2 - X$$

and the speed of the particle A is

$$\frac{dX}{dt} = \frac{d}{dt}(\tfrac{1}{2}ft^2 - x) = ft - \frac{dx}{dt}$$

From the given information the equation of motion of the particle in the direction AB is

$$m \frac{\mathrm{d}^2 X}{\mathrm{d}t^2} = \text{(tension in string)} - \text{(resisting force)}$$

Now:

$$\frac{\mathrm{d}^2 X}{\mathrm{d}t^2} = f - \frac{\mathrm{d}^2 x}{\mathrm{d}t^2}$$

$$\text{tension in string} = \frac{3lmn^2}{l} x$$

$$\text{resisting force} = 4mnv$$

$$= 4mn \frac{\mathrm{d}X}{\mathrm{d}t}$$

$$= 4mn \left[ft - \frac{\mathrm{d}x}{\mathrm{d}t} \right]$$

So:

$$m \left(f - \frac{\mathrm{d}^2 x}{\mathrm{d}t^2} \right) = 3mn^2 x - 4mn \left(ft - \frac{\mathrm{d}x}{\mathrm{d}t} \right)$$

or:

$$\frac{\mathrm{d}^2 x}{\mathrm{d}t^2} + 4n \frac{\mathrm{d}x}{\mathrm{d}t} + 3n^2 x = 4nft + f \qquad (1)$$

The complementary equation is

$$\ddot{x} + 4n\dot{x} + 3n^2 x = 0$$

with auxiliary equation

$$\lambda^2 + 4n\lambda + 3n^2 = 0$$

This equation factorises to give:

$$(\lambda + 3n)(\lambda + n) = 0$$

and so has roots $\lambda = -3n$ and $\lambda = -n$.

Hence the complementary function is

$$x_c = A\mathrm{e}^{-3nt} + B\mathrm{e}^{-nt}$$

where A and B are arbitrary constants.

To find a particular solution try:

$$x_p = at + b$$

where a and b are constants to be determined. Using $\dot{x}_p = a$ and $\ddot{x}_p = 0$ and substituting into equation (1) gives:

$$0 + 4na + 3n^2(at + b) = 4nft + f$$

If you equate the terms involving t on the two sides of this equation, you get:

$$3n^2at = 4nft$$

so:

$$a = \frac{4f}{3n}$$

From the terms not involving t:

$$4na + 3n^2b = f$$

Substituting the value of a found above gives:

$$\frac{16f}{3} + 3n^2b = f$$

$\Rightarrow$

$$b = \frac{1}{3n^2}\left(f - \frac{16f}{3}\right) = -\frac{13f}{9n^2}$$

So:

$$x_p = \left(\frac{4f}{3n}\right)t - \frac{13f}{9n^2}$$

and the general solution is

$$x = Ae^{-3nt} + Be^{-nt} + \left(\frac{4f}{3n}\right)t - \frac{13f}{9n^2}$$

Using the initial conditions $x = 0$ and $\dot{x} = 0$ when $t = 0$ gives:

$x = 0$:

$$A + B - \frac{13f}{9n^2} = 0$$

$\dot{x} = 0$:

$$-3nA - nB + \frac{4f}{3n} = 0$$

Solving these equations simultaneously for A and B gives:

$$A = -\frac{f}{18n^2} \qquad B = \frac{3f}{2n^2}$$

and so:

$$x = \left(-\frac{f}{18n^2}\right)e^{-3nt} + \left(\frac{3f}{2n^2}\right)e^{-nt} + \left(\frac{4f}{3n}\right)t - \frac{13f}{9n^2}$$

Exercise 4A

In questions 1–3 the differential equation is the equation of motion of a particle P moving in a straight line under the action of two forces:

(i) a restoring force directed towards a fixed point O of the line with magnitude proportional to the distance x m of P from O,

(ii) a resistance to its motion with magnitude proportional to its speed.

1 $\ddot{x} + 2\dot{x} + 2x = 0$ with $x = 1$ and $\dot{x} = 0$ when $t = 0$. Calculate x when $t = \dfrac{\pi}{2}$ and describe the motion.

2 $\ddot{x} + 2\dot{x} + x = 0$ with $x = 4$ and $\dot{x} = 0$ when $t = 0$. Find x as a function of t and show that x never vanishes for $t \geqslant 0$. Find the speed of P when $t = 2$.

3 $\ddot{x} + 2\dot{x} + 5x = 0$ with $x = 2$ and $\dot{x} = 0$ when $t = 0$. Find x as a function of t and show that the smallest non-zero value of t for which $\dot{x}$ is zero is $t = \dfrac{\pi}{2}$.

4 A particle P of mass 1 kg moves in a horizontal straight line under the action of a force directed towards a fixed point O of the line. The force varies as the distance of the particle from O and is equal to $2x$ N when P is at a distance x metres from O. The particle is also subject to a resisting force whose magnitude is proportional to its speed and which is equal to $3v$ N when the speed of P is $v\,\mathrm{m\,s^{-1}}$. Given that P starts from rest when $x = 1$ and $t = 0$, find x as a function of t and obtain the value of t when $x = \frac{1}{10}$.

5 A particle P moves along a straight line so that its displacement, x metres, from a fixed point O of the line at time t seconds is given by

$$x = 3\mathrm{e}^{-t} \cos 2t$$

(a) Show that the particle performs oscillations about O with constant period.
(b) Show that the amplitude of the motion decreases in geometric progression.
(c) Show that $\ddot{x} + 2\dot{x} + 5x = 0$.

6 A particle of mass m moves in a straight line. At time t its displacement from a fixed point O of the line is x. Explain the nature of the forces acting on the particle, given that the equation of motion can be written as:

$$\frac{\mathrm{d}^2 x}{\mathrm{d}t^2} + 2k\frac{\mathrm{d}x}{\mathrm{d}t} + 10k^2 x = 0$$

where k is a positive constant.

Given that $x = 0$ and $\dfrac{dx}{dt} = u$ when $t = 0$,

(a) find x as a function of t

(b) show that when x is next zero

$$\frac{dx}{dt} = -ue^{-\pi/3}$$

7 A particle P of mass m is attached to one end of a light spring of modulus $10mn^2a$ and natural length a. The other end of the spring is attached to a fixed point O. The particle rests in equilibrium at the point A directly below O. At time $t = 0$ an impulse is applied to P so that it starts to move downwards with speed V. The motion of P is also subject to a resistance of magnitude $2mnv$ where v is the speed of P. If x is the displacement of P from A at time t

(a) show that $\quad \dfrac{d^2x}{dt^2} + 2n\dfrac{dx}{dt} + 10n^2x = 0$

(b) Find x as a function of t and sketch a graph of this function.

(c) Show that the particle is instantaneously at rest when $nt = \frac{1}{3}k\pi + \alpha$ where k is a non-negative integer and α is the acute angle such that $\tan 3\alpha = 3$.

8 A particle P is attached to one end of a spring, the other end of which oscillates. The displacement x of P from a fixed reference point at time t satisfies the differential equation

$$\frac{d^2x}{dt^2} + 4x = \lambda \sin t$$

where λ is a constant. Given that $x = 0$ and $\dfrac{dx}{dt} = \dfrac{2\lambda}{3}$ when $t = 0$, find x as a function of t.

9 A light spring AB, having natural length a and modulus of elasticity $3amn^2$, lies straight and at its natural length at rest on a horizontal table. A particle of mass m is attached to the end A. The end B is then moved in a straight line in the direction AB with constant speed V. The resulting motion of the particle is resisted by a force of magnitude $4mnv$ where v is the speed of the particle. If x is the extension of the spring at time t, show that

$$\frac{d^2x}{dt^2} + 4n\frac{dx}{dt} + 3n^2x = 4nV$$

Obtain x in terms of t.

10 A particle of mass m lies at rest on a horizontal table and is attached to one end of a light spring which, when stretched, exerts a tension of magnitude $m\omega^2 \times$ (extension of the spring), where ω is constant. The other end of the spring is now moved with constant speed u along the table in a direction away from the particle. The resistance to the motion of the particle provided by the table is of magnitude $mk \times$ (the speed of the particle), where k is constant.

(a) Obtain the differential equation satisfied by x, the extension of the spring after time t.

(b) Given that $k = 2\omega$ show that

$$x = \frac{u}{\omega}[2 - (2 + \omega t)e^{-\omega t}]$$

11

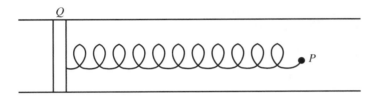

The figure shows a particle P of mass 1 kg which is free to slide horizontally inside a smooth cylindrical tube. The particle is attached to one end of a light elastic spring of natural length 0.5 m and modulus of elasticity 2 N. The system is initially at rest. The other end Q of the spring is then forced to oscillate with simple harmonic motion so that at time t seconds its displacement from its initial position is $\frac{1}{4}\sin 3t$ metres. The displacement of P from its initial position at time t seconds is x metres, measured in the same direction as the displacement of Q.

(a) Show that $\dfrac{\mathrm{d}^2 x}{\mathrm{d}t^2} + 4x = \sin 3t$.

(b) Find the first time, after the motion starts, at which P is instantaneously at rest.

12 A light elastic string of natural length l and modulus $4mln^2$ has a particle of mass m attached to its mid-point. One end of the string is attached to a fixed point A on a smooth horizontal table and the other end to a fixed point B on the same table where $AB = 2l$. The particle moves along the line AB and is subject to a resistance of magnitude $2mnv$, where v is the speed of the particle. The mid-point of AB is O.

(a) Show that, so long as both parts of the string remain taut, the displacement x of P from O at time t satisfies the differential equation

$$\ddot{x} + 2n\dot{x} + 16n^2 x = 0$$

Given that initially P starts from rest when $x = \dfrac{l}{4}$
(b) find x as a function of t.

SUMMARY OF KEY POINTS

1 When resistance, proportional to the speed, is taken into account the S.H.M. equation becomes

$$\ddot{x} + k\dot{x} + \omega^2 x = 0$$

The form of the solution depends on the nature of the solution of the corresponding auxiliary equation

$$\lambda^2 + k\lambda + \omega^2 = 0$$

that is, whether the roots are real and distinct, equal or complex.

2 When there is an additional time-dependent force $mf(t)$ the damped harmonic equation becomes

$$\ddot{x} + k\dot{x} + \omega^2 x = f(t)$$

The general solution of this equation is

$$x = x_c \text{ (complementary function)}$$
$$+ x_p \text{ (particular integral)}$$

Stability

5

5.1 Equilibrium and the potential energy test for stability

In Book M2 section 5.2, it was seen that a rigid body moving in a plane is in equilibrium if:

(i) the vector sum of the forces acting is zero
(ii) the algebraic sum of the moments of the forces, about any given point, is zero.

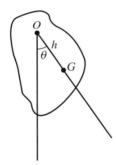

Consider the body of mass M, shown in the diagram. The body is free to rotate about a smooth horizontal axis through some point O of the body. Let the distance of the centre of mass G from the axis be h. In a general position, OG makes an angle θ with the downward vertical.

There are two positions of equilibrium:

(a) with G vertically below O,   (b) with G vertically above O.

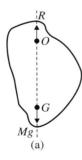

(a)

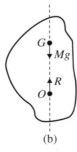

(b)

In both cases all the forces acting on the body are vertical and act through the axis at O.

However it is clear that there is a distinct difference between these two cases. In case (a), if the body is slightly disturbed from this position it will oscillate about its equilibrium position. In case (b), if the body is slightly disturbed it will move right away from its position of equilibrium.

Case (a) is a case of **stable equilibrium** and case (b) is a case of **unstable equilibrium**.

■ **A position of equilibrium for any system is said to be stable when an arbitrary small disturbance does not cause the system to depart from the position of equilibrium. Otherwise it is unstable.**

Relative to the fixed level through O, the potential energy of the body, in the general position shown earlier, is:

$$V = -Mgh\cos\theta$$

Then:
$$\frac{\mathrm{d}V}{\mathrm{d}\theta} = Mgh\sin\theta$$

And:
$$\frac{\mathrm{d}^2V}{\mathrm{d}\theta^2} = Mgh\cos\theta$$

Hence V has stationary values when $\dfrac{\mathrm{d}V}{\mathrm{d}\theta} = 0$, that is when $\sin\theta = 0$.

So $\theta = 0$ and $\theta = \pi$ give stationary values for V. Further, when $\theta = 0$, $\dfrac{\mathrm{d}^2V}{\mathrm{d}\theta^2} = Mgh$, which is greater than zero, and so V has a minimum for this value of θ.

When $\theta = \pi$, $\dfrac{\mathrm{d}^2V}{\mathrm{d}\theta^2} = -Mgh$, which is negative, and so V has a maximum for this value of θ.

The graph of V against θ shows all these properties of V clearly.

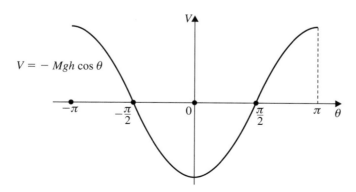

- *V* is a **minimum** at the position of **stable equilibrium.**
  V is a **maximum** at the position of **unstable equilibrium.**

As a further illustration consider a bead moving on a smooth wire which is fixed in a vertical plane as shown in the diagram.

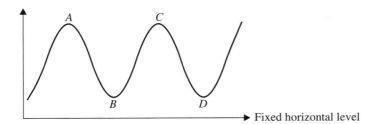

The only points at which the bead can rest in equilibrium are the points *A*, *B*, *C* and *D* where the reaction due to the wire is vertical. Since the potential energy of the bead is directly proportional to the height of the particle above a fixed horizontal level, these equilibrium positions coincide with the positions where *V* is stationary. So:
V is a **maximum** at *A* and *C*, which are **unstable** positions of equilibrium.
V is a **minimum** at *B* and *D*, which are **stable** positions of equilibrium.

These two examples illustrate the following general results:

- **The energy condition for equilibrium**
  In a mechanical system which is free to move, and to whose motion the conservation of energy can be applied, possible positions of equilibrium occur where the potential energy of the system has stationary values.
- **Energy criterion for stability**
  In a mechanical system to which the energy condition for equilibrium applies, **minimum** values of the potential energy correspond to positions of **stable equilibrium** and **maximum** values correspond to positions of **unstable equilibrium.**

Example 1
A uniform rod *AB*, of mass 12*m* and length 2*a*, can turn freely about the end *A* which is smoothly hinged to a vertical wall. A light inextensible string is attached to the other end, *B*, of the rod and passes through a small smooth ring fixed at the point *C*, at a distance 2*a* from *A* and at the same level as *A*. To the other end of the string is attached a particle *P* of mass *m*. Show that the system is in stable equilibrium when $8\cos\theta = 1$, where θ is the inclination of *AB* to the horizontal.

The information given is shown in the diagram.

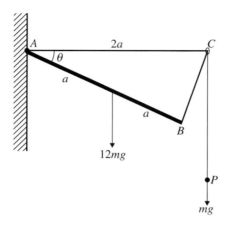

The potential energy of the rod, taking AC as the zero level, is $-12mga \sin\theta$. The potential energy of the particle P is $-mg PC$.

Let the length of the string BP be l.

Then:
$$PC = l - BC$$

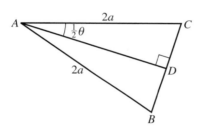

The triangle ABC is isosceles since $AC = AB = 2a$. Therefore, if D is the mid-point of BC, then:

$$\angle ADC = \frac{\pi}{2}$$

And:
$$\angle DAC = \frac{\theta}{2}$$

From triangle ADC:
$$CD = 2a \sin\frac{\theta}{2}$$

So:
$$BC = 2CD = 4a \sin\frac{\theta}{2}$$

The potential energy of the particle P is $-mg\left(l - 4a\sin\frac{\theta}{2}\right)$

So the total potential energy of the system is:

$$V = -12mga \sin\theta + 4mga \sin\frac{\theta}{2} - mgl$$

So:
$$\frac{dV}{d\theta} = -12mga\cos\theta + 4mga\left(\tfrac{1}{2}\cos\frac{\theta}{2}\right)$$

$$= -12mga\cos\theta + 2mga\cos\frac{\theta}{2} \qquad (1)$$

For stationary values $\dfrac{dV}{d\theta} = 0$

So:
$$-12\cos\theta + 2\cos\frac{\theta}{2} = 0$$

Defining $c = \cos\dfrac{\theta}{2}$, and using $\cos\theta = 2\cos^2\left(\dfrac{\theta}{2}\right) - 1$, gives:

$$-12(2c^2 - 1) + 2c = 0$$

Or:
$$-2(12c^2 - c - 6) = 0$$

But:
$$12c^2 - c - 6 = (4c - 3)(3c + 2)$$

So: $\dfrac{dV}{d\theta} = 0$ when $c = \tfrac{3}{4}$ and $c = -\tfrac{2}{3}$.

When $c = \tfrac{3}{4}$ then:

$$\cos\theta = 2\left(\frac{3}{4}\right)^2 - 1$$

$$= 2\left(\frac{9}{16}\right) - 1 = \frac{1}{8}$$

and so $8\cos\theta = 1$.

Differentiating equation (1) gives:

$$\frac{d^2V}{d\theta^2} = 12mga\sin\theta + 2mga\left(-\tfrac{1}{2}\sin\frac{\theta}{2}\right)$$

$$= 24mga\sin\frac{\theta}{2}\cos\frac{\theta}{2} - mga\sin\frac{\theta}{2}$$

using $\sin\theta = 2\sin\dfrac{\theta}{2}\cos\dfrac{\theta}{2}$.

If $\cos\dfrac{\theta}{2} = \tfrac{3}{4}$ then by Pythagoras' theorem:

$$\sin^2\frac{\theta}{2} = 1 - \cos^2\frac{\theta}{2} = 1 - \frac{9}{16} = \frac{7}{16}$$

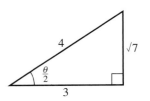

So:
$$\sin\frac{\theta}{2} = \frac{\sqrt{7}}{4}$$

Hence when $\cos\theta = \frac{1}{8}$:

$$\frac{d^2V}{d\theta^2} = 24\,mga\left(\frac{\sqrt{7}}{4}\right)\left(\frac{3}{4}\right) - mga\left(\frac{\sqrt{7}}{4}\right)$$

$$= mga\,\frac{17}{4}\sqrt{7} - mga\,\frac{\sqrt{7}}{4}$$

$$= mga\,\frac{18}{4}\sqrt{7}$$

Since $\dfrac{d^2V}{d\theta^2}$ is greater than zero, this position of equilibrium is stable.

Example 2

A particle P, of mass m, is attached to one end of an elastic string of natural length l and modulus of elasticity λ. The other end of the string is attached to a fixed point O. Obtain the potential energy V of the system, when the extension of the string is x. Hence obtain the value of x when the system is in equilibrium and draw a sketch of V against x.

The potential energy of the particle P is $-mg(l+x)$, taking O as the level of zero.

The elastic potential energy of the string is $\dfrac{\lambda x^2}{2l}$. (See Book M3 section 2.2.)

Hence the total potential energy is:

$$V = \frac{\lambda x^2}{2l} - mgx + \text{constant}$$

The system is in equilibrium when $\dfrac{dV}{dx} = 0$.

That is:

$$\frac{\lambda}{2l}(2x) - mg = 0$$

Or:

$$x = \frac{mgl}{\lambda}$$

As $\dfrac{d^2V}{dx^2} = \dfrac{\lambda}{l}$, which is greater than zero, then V is a minimum.

The graph of V against x is:

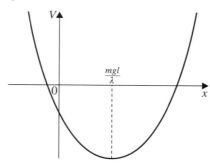

Example 3

A smooth ring P, of mass m, is free to slide on a smooth fixed vertical circular wire of radius a and centre O. A light elastic string of natural length $2a$ and modulus kmg passes through the ring, its ends being fixed to points A and B which are the ends of the horizontal diameter of the wire. Show that, when OP makes an angle 2θ with AB, the ring being below AB, the potential energy of the system is given by:

$$V = kmga(\sin\theta + \cos\theta - 1)^2 - mga\sin 2\theta + \text{constant}$$

Show that if $k > 2 + \sqrt{2}$, the ring is in unstable equilibrium when at the lowest point of the wire.

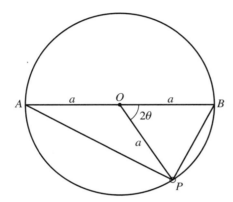

As in Example 1, the triangle OBP is isosceles with $OB = OP$ and therefore the length of BP is $2a\sin\theta$.

Similarly, since the angle $AOP = \pi - 2\theta$ then:

$$\text{length of } AP = 2a\sin\left(\frac{\pi}{2} - \theta\right) = 2a\cos\theta$$

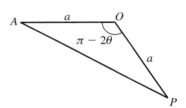

Hence the total length of the elastic string in this position is:

$$AP + PB = 2a\cos\theta + 2a\sin\theta$$

The elastic potential energy of the string is then:

$$\text{P.E.} = \frac{kmg}{2(2a)}[2a\cos\theta + 2a\sin\theta - 2a]^2$$

The potential energy of the particle P, taking AB as the zero level, is:

$$\text{P.E.} = -mga\sin 2\theta$$

So the total potential energy of the system is:

$$V = kmga(\sin\theta + \cos\theta - 1)^2 - mga\sin 2\theta + \text{constant}$$

The positions of equilibrium are given by $\dfrac{dV}{d\theta} = 0$, that is:

$$\frac{dV}{d\theta} = 2kmga(\sin\theta + \cos\theta - 1)(\cos\theta - \sin\theta) - mga\,2\cos 2\theta = 0$$

Since $\cos 2\theta = \cos^2\theta - \sin^2\theta$ this gives:

$$\frac{dV}{d\theta} = 2mga\,(\cos\theta - \sin\theta)[(k-1)(\sin\theta + \cos\theta) - k] = 0$$

One solution of this equation is given by:

$$\cos\theta = \sin\theta$$

or:

$$\tan\theta = 1$$

$$\Rightarrow \qquad \theta = \frac{\pi}{4}$$

In this case the ring is at the lowest point as $2\theta = \frac{\pi}{2}$.

To investigate the nature of this position, $\dfrac{d^2V}{d\theta^2}$ is required:

$$\frac{d^2V}{d\theta^2} = 2mga\{(\cos\theta - \sin\theta)(k-1)(\cos\theta - \sin\theta)$$
$$+ [(k-1)(\sin\theta + \cos\theta) - k](-\sin\theta - \cos\theta)\}$$

Substituting $\theta = \dfrac{\pi}{4}$ and using $\sin\dfrac{\pi}{4} = \cos\dfrac{\pi}{4} = \dfrac{1}{\sqrt{2}}$ gives:

$$\frac{d^2V}{d\theta^2} = 2mga\left[(k-1)\left(\frac{1}{\sqrt{2}} + \frac{1}{\sqrt{2}}\right) - k\right]\left(-\frac{1}{\sqrt{2}} - \frac{1}{\sqrt{2}}\right)$$

$$= 2mga\left[-2(k-1) + \frac{2k}{\sqrt{2}}\right]$$

$$= 2mga\left[2k\left(\frac{1}{\sqrt{2}} - 1\right) + 2\right]$$

This position of equilibrium is unstable if $\dfrac{d^2V}{d\theta^2}$ is less than zero, that is if:

$$2k\left(\frac{1}{\sqrt{2}} - 1\right) + 2 < 0$$

or:

$$2k\left(1 - \frac{1}{\sqrt{2}}\right) > 2$$

So:

$$k > \frac{\sqrt{2}}{\sqrt{2} - 1} = \frac{\sqrt{2}(\sqrt{2}+1)}{(\sqrt{2}+1)(\sqrt{2}-1)} = 2 + \sqrt{2}$$

Example 4

A uniform lamina, of mass M, is in the shape of an isosceles triangle ABC with $AB = AC$. The mid-point of BC is D and $AD = 4l$. The lamina can turn freely about the smooth axis BC which is fixed and horizontal. One end of a light elastic string of natural length l and modulus $\dfrac{5Mg}{24}$ is attached to the vertex A of the triangle. The other end of the string is attached to a point P at a height $4l$ vertically above D.

(a) Obtain the potential energy of the system when DA makes an angle 2θ with the upward vertical.
(b) Find the positions of equilibrium and determine their nature.

The diagram shows the plane PAD, the smooth axis BC intersecting this plane in D.

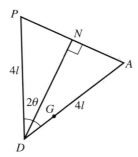

As before, triangle PAD is isosceles with $DP = DA$ and therefore, if N is the mid-point of AP, then $\angle DNA = \dfrac{\pi}{2}$ and $\angle NDA = \theta$.

Hence $NA = 4l \sin \theta$ and $AP = 2NA = 8l \sin \theta$.

The elastic potential energy of the elastic string AP is then:

$$\frac{5Mg}{24} \frac{1}{2l} (8l \sin \theta - l)^2$$

From Book M2 section 2.2, the centre of mass of the triangular lamina is at a distance of $\frac{1}{3}DA$ from D. So $DG = \frac{1}{3}4l$.

Hence the potential energy of the lamina, taking the level through D as the zero level is:

$$Mg \frac{4l}{3} \cos 2\theta$$

So the total potential energy of the system is:

$$
\begin{aligned}
V &= \frac{4Mgl}{3} \cos 2\theta + \frac{5Mg}{48l} (8l \sin \theta - l)^2 \\
&= \frac{4Mgl}{3} (1 - 2\sin^2 \theta) + \frac{5Mg}{48l} (64l^2 \sin^2 \theta - 16l^2 \sin \theta + l^2) \\
&= -\frac{8Mgl}{3} \sin^2 \theta + \frac{5Mgl}{48} (64 \sin^2 \theta) - \frac{5Mgl}{48} (16 \sin \theta) + \text{constant} \\
&= \frac{Mgl}{3} (12 \sin^2 \theta - 5 \sin \theta) + \text{constant}
\end{aligned}
$$

So: $\qquad \dfrac{\mathrm{d}V}{\mathrm{d}\theta} = \dfrac{Mgl}{3} (24 \sin \theta \cos \theta - 5 \cos \theta)$

And: $\qquad \dfrac{\mathrm{d}^2 V}{\mathrm{d}\theta^2} = \dfrac{Mgl}{3} (24 \cos^2 \theta - 24 \sin^2 \theta + 5 \sin \theta)$

The positions of equilibrium are given by $\dfrac{dV}{d\theta} = 0$.

So:
$$24 \sin\theta\cos\theta - 5\cos\theta = 0$$

or:
$$\cos\theta\,(24\sin\theta - 5) = 0$$

So: (i) $\cos\theta = 0$

or (ii) $24\sin\theta - 5 = 0$
$$\sin\theta = \tfrac{5}{24}$$

(i) $\cos\theta = 0$

$\Rightarrow \quad \theta = \dfrac{\pi}{2}$

In this position, $2\theta = \pi$ and A is vertically below D.

When $\theta = \dfrac{\pi}{2}$,

$$\frac{d^2V}{d\theta^2} = \frac{Mgl}{3}(-24 + 5) = -19\frac{Mgl}{3}$$

which is negative and so this position is **unstable**.

(ii) $\sin\theta = \tfrac{5}{24}$

$\Rightarrow \quad \theta = 0.21 \quad$ or $\quad \theta = 2.93°$

$\Rightarrow \quad 2\theta = 0.42 \quad$ or $2\theta = 5.86°$

Then:
$$\frac{d^2V}{d\theta^2} = \frac{Mgl}{3}[24(1 - \sin^2\theta) - 24\sin^2\theta + 5\sin\theta]$$

$$= \frac{Mgl}{3}(24 - 48\sin^2\theta + 5\sin\theta)$$

$$= \frac{Mgl}{3}\left(24 - \frac{48 \times 25}{24 \times 24} + \frac{5 \times 5}{24}\right) = \frac{551}{72}Mgl$$

So $\dfrac{d^2V}{d\theta^2}$ is greater than zero and these positions are **stable**.

Exercise 5A

1 A uniform rod, of mass m and length $2a$, is smoothly hinged at one end to a fixed point A. The other end of the rod is attached to a light inextensible string which passes over a smooth peg B at the same level as A and at a distance $2a$ from it. A mass $\dfrac{m\sqrt{3}}{6}$ is attached to the other end of the string.

Express the potential energy V of the system in terms of θ, the angle which the rod makes with the line AB.

Show that there is a position of equilibrium when $\theta = \dfrac{\pi}{3}$, and determine whether this position of equilibrium is stable or unstable.

2 A uniform rod AB, of mass m and length $2a$, is free to rotate in a vertical plane about the end A. A light elastic string, of modulus kmg and natural length a, has one end attached to B and the other end to a fixed point O which is vertically above A with $OA = 2a$. Show that when AB makes an angle θ with the downward vertical, the potential energy V of the system, when the string is stretched, is given by

$$V = mga\left[(4k - 1)\cos\theta - 4k\cos\left(\frac{\theta}{2}\right)\right] + \text{constant}$$

Deduce that, if $k > \frac{1}{3}$, the equilibrium position in which the rod is vertical, with B below A, is unstable, and that there is an oblique position of equilibrium which is stable.

3 A uniform heavy rod AB, of mass m and length $4a$, can turn in a vertical plane about one end A which is fixed. To the other end B is attached a light elastic string of natural length $3a$ and modulus $\frac{1}{2}mg$. The other end of the string is attached to a light ring which can slide on a smooth horizontal bar which is fixed at a height of $8a$ above A and in the vertical plane through AB. Find the equilibrium positions of the rod and determine their nature.

4 A uniform rod AB, of length $2a$ and mass $2m$, rests with one end A in contact with a smooth vertical wall so that the rod lies in a plane perpendicular to the wall. The rod is supported by a smooth fixed horizontal rail which is parallel to the wall and at a distance c from it. A particle of mass m is attached to the end B of the rod. Show that when the rod makes an angle θ with the upward vertical the potential energy of the system is

$$mg(4a\cos\theta - 3c\cot\theta) + \text{constant}$$

Show also that if $3c > 4a$ there is no equilibrium position, but if $3c < 4a$ there is one equilibrium position. Determine whether this position is stable or unstable.

5 A bead of mass m can slide on a smooth circular hoop of wire of radius a which is fixed in a vertical position. One end of a light spring, of natural length a and modulus $3mg$, is attached to the bead. The other end of the spring is attached to the highest point of the wire. Show that, when the spring makes an angle θ with the downward vertical, the potential energy V of the system is given by

$$V = \tfrac{3}{2}mga\,(2\cos\theta - 1)^2 - 2mga\cos^2\theta + \text{constant}$$

Show also that possible equilibrium positions occur when $\theta = 0$ and $\cos\theta = \tfrac{3}{4}$. Determine the nature of these equilibrium positions.

6 Four uniform rods, each of mass m and length $2a$, are smoothly jointed together to form a rhombus $ABCD$. A light elastic string of modulus $2mg$ and natural length a connects A and C. The vertex A is smoothly pinned to a fixed support and the system hangs at rest. Show that there is a position of stable equilibrium in which the angle BCD of the rhombus is $\tfrac{2}{3}\pi$.

7 A bead P, of mass m, can slide on a smooth wire which is bent into a semicircle of radius a and is fixed with its plane horizontal. Two identical elastic strings of natural length $l(< a\sqrt{2})$ and modulus of elasticity λ, have one end attached to the bead. The other ends are attached to the ends A and B of the wire. Show that the symmetrical position of the system is stable.

8 A small bead B of mass m can slide on a smooth circular wire of radius a which is fixed in a vertical plane. B is attached to one end of a light elastic string, of natural length $\tfrac{3}{2}a$ and modulus of elasticity mg. The other end of the string is attached to a fixed point A which is vertically above the centre C of the circular wire with $AC = 3a$.
Show that there is a stable equilibrium position of the system when θ, the angle which the radius through B makes with the downward vertical, satisfies $\cos\theta = -\tfrac{1}{6}$.

9 A uniform rod AB, of mass m and length $2a$, is free to rotate in a vertical plane about the point A. A light elastic string of natural length a and modulus kmg has one end attached to B and the other end attached to a fixed point O, which is vertically above A, with $AO = 2a$.

(a) Show that when AB makes an angle θ with the downward vertical the potential energy V of the system, when the string is stretched, is given by

$$V = mga\left[(4k-1)\cos\theta - 4k\cos\left(\frac{\theta}{2}\right)\right] + \text{constant}$$

Given that $k > \frac{1}{3}$,

(b) show that the equilibrium position in which the rod is vertical with B below A is unstable.

(c) show that there is an oblique position of equilibrium which is stable.

SUMMARY OF KEY POINTS

1 In a mechanical system which is free to move, and to whose motion the conservation of energy can be applied, possible positions of equilibrium occur when the potential energy of the system has stationary values.

2 In a mechanical system to which the energy condition for equilibrium applies:
   - **minimum** values of the potential energy correspond to positions of **stable equilibrium**
   - **maximum** values of the potential energy correspond to positions of **unstable equilibrium**.

Review exercise 2

1 A particle P of mass m moves in a medium which produces a resistance of magnitude mkv, where v is the speed of P and k is a constant. The particle P is projected vertically upwards in this medium with speed $\frac{g}{k}$. Show that P comes momentarily to rest after time $\frac{\ln 2}{k}$.

Find, in terms of k and g, the greatest height above the point of projection reached by P.　　　　　[E]

2 A particle of mass m is projected vertically upwards with speed u in a medium which exerts a resisting force of magnitude mkv, where v is the speed of the particle and k is a positive constant. Find the time taken to reach the highest point, and show that the greatest height attained above the point of projection is

$$\frac{1}{k^2}\left[uk - g\ln\left(1 + \frac{ku}{g}\right)\right]$$

Find, in terms of k, g and T, the speed of the particle at time T after it has reached its greatest height, and hence, or otherwise, show that this speed tends to a finite limit as T increases indefinitely.　　　　　[E]

3 A particle of mass m moves in a straight line on a horizontal table against a resistance of magnitude $\lambda(mv + k)$, where v is the speed and λ and k are positive constants. Given that the particle starts with speed u at time $t = 0$, show that the speed v of the particle at time t is

$$v = \frac{k}{m}(e^{-\lambda t} - 1) + ue^{-\lambda t}$$

　　　　　[E]

4 A particle, of mass m, moves under gravity down a line of greatest slope of a smooth plane inclined at an angle α to the horizontal. When the speed of the particle is v, the resistance to the motion of the particle is mkv, where k is a positive constant. Show that the limiting speed c of the particle is given by

$$c = \frac{g \sin \alpha}{k}$$

The particle starts from rest. Show that the time T taken to reach a speed of $\frac{1}{2}c$ is given by

$$T = \frac{1}{k} \ln 2$$

Find, in terms of c and k, the distance travelled by the particle in attaining the speed of $\frac{1}{2}c$.

5 A ship, of mass m, is propelled in a straight line through the water by a propeller which develops a constant force of magnitude F. When the speed of the ship is v, the water causes a drag, of magnitude kv, where k is a constant, to act on the ship. The ship starts from rest at time $t = 0$. Show that the ship reaches half its theoretical maximum speed of $\frac{F}{k}$ when

$$t = \frac{(m \ln 2)}{k}.$$

When the ship is moving with speed $\frac{F}{2k}$, an emergency occurs and the captain reverses the engines so that the propeller force, which remains of magnitude F, acts backwards. Show that the ship covers a further distance

$$\frac{mF}{k^2} \left[\frac{1}{2} - \ln \left(\frac{3}{2} \right) \right]$$

on its original course, which may be assumed to remain unchanged, before being brought to rest. [E]

6 A particle is projected vertically upwards with speed U in a medium in which the resistance to motion is proportional to the square of the speed. Given that U is also the speed for which the resistance offered by the medium is equal to the weight of the particle, show that the time of ascent is $\frac{\pi U}{4g}$ and that the distance ascended is $\frac{U^2}{2g} \ln 2$. [E]

7 A particle P of mass m moves under gravity in a medium which is such that the resistance to motion is of magnitude mkv^2, where v is the speed of P and k is a positive constant. Show that it is possible for P to fall vertically with a constant speed

$$U = \sqrt{\left(\frac{g}{k}\right)}.$$

Given that P is projected vertically upwards with speed $V(>U)$, show that the speed of P is equal to U when the height of P above the point of projection is

$$\frac{U^2}{2g}\ln\left(\frac{V^2 + U^2}{2U^2}\right)$$

Find, in terms of U, V and g, the time taken for the speed of P to decrease from V to U. [E]

8 At time t, a particle P, of mass m, moving in a straight line has speed v. The only force acting is a resistance of magnitude $mk(V_0^2 + 2v^2)$, where k is a positive constant and V_0 is the speed of P when $t = 0$. Show that, as v reduces from V_0 to $\frac{V_0}{2}$, P travels a distance $\frac{\ln 2}{4k}$.

Express the time P takes to cover this distance in the form $\frac{\lambda}{kV_0}$, giving the value of λ to two decimal places. [E]

9 A train, total mass M, including the engine, is moving along a straight horizontal track. The engine exerts a constant driving force of magnitude F. At any instant the total resistance is bv^2 where b is a positive constant and v is the speed of the train at that instant. Show that the limiting speed V of the train is $\left(\frac{F}{b}\right)^{\frac{1}{2}}$.

The train starts from rest at $t = 0$. Show that it reaches a speed of $\frac{1}{2}V$ after a time

$$\frac{M}{2bV}\ln 3$$

Show further that the distance covered in this time is

$$\frac{M}{2b}\ln\left(\frac{4}{3}\right)$$ [E]

10   The equation of motion of a particle moving on the x-axis is

$$\ddot{x} + 2k\dot{x} + n^2 x = 0$$

Given that k and n are positive constants, with $k < n$, show that the time between two successive maxima of $|x|$ is constant.   [E]

11   A particle of mass m is constrained to move along the horizontal straight line Ox. When the particle is at P, where the displacement of P from O is x, the particle is subject to a force of magnitude $m\omega^2|x|$ acting towards O, where ω is a constant. The particle is also subject to a resistance of magnitude $2mk|\dot{x}|$, where k is a positive constant with $k < \omega$. Show that, at time t,

$$x = e^{-kt}(A\cos nt + B\sin nt)$$

where A and B are arbitrary constants and $n^2 = \omega^2 - k^2$. Explain why the motion can be regarded as an oscillation about O with period $2\pi/n$ but with decreasing amplitude. Show that the amplitude decreases by a factor $e^{-2\pi k/n}$ in one complete oscillation.
In time T, a particle performs an exact number of complete oscillations. The final amplitude is one third of the initial amplitude. Show that

$$kT = \ln 3$$ [E]

12   Show that the roots of the equation

$$\lambda^2 + k\lambda + \omega^2 = 0$$

are distinct and both negative when $k > 2\omega > 0$.
A particle moves along the x-axis under the action of a force $\omega^2|x|$ per unit mass directed towards the origin and a resisting force kv per unit mass, where k and ω are positive constants and x, v are the displacement from the origin and speed respectively after time t. Show that the differential equation satisfied by x and t is

$$\frac{d^2x}{dt^2} + k\frac{dx}{dt} + \omega^2 x = 0$$

The particle starts from rest when $x = a$. Show that, if $k > 2\omega > 0$,

$$(\lambda_2 - \lambda_1)x = a(\lambda_2 e^{-\lambda_1 t} - \lambda_1 e^{-\lambda_2 t})$$

where $-\lambda_1, -\lambda_2$ are the roots of the quadratic equation

$$\lambda^2 + k\lambda + \omega^2 = 0$$

Further, show that the particle does not pass through the origin. [E]

13 A particle moves on the Ox axis and its displacement x at time t is governed by the equation

$$\frac{d^2 x}{dt^2} + 2k\frac{dx}{dt} + n^2 x = 0$$

where k and n are positive constants. Given that $x = a$ and $\frac{dx}{dt} = 0$ when $t = 0$, find x in each of the two cases

(a) $k^2 = 2n^2$
(b) $k^2 = \frac{3}{4}n^2$

Show that in case (b) the time interval between successive stationary values of x is constant. [E]

14 A particle P of mass m is suspended from a fixed point by a spring of natural length l and modulus $2mn^2 l$. The particle is projected vertically downwards with speed V from its equilibrium position. The motion of the particle is resisted by a force of magnitude $2mn$ times its speed acting in a direction opposite to its motion. Given that x is the displacement of P downwards from the equilibrium position at time t, show that

$$\frac{d^2 x}{dt^2} + 2n\frac{dx}{dt} + 2n^2 x = 0$$

Find x in terms of t and sketch the graph of x against t. Show that P is instantaneously at rest when $nt = (k + \frac{1}{4})\pi$, where $k \in \mathbb{N}$. [E]

15 A particle P of mass m is suspended from a fixed point by a light elastic string of natural length a and modulus $k^2 mg$, where k is a positive constant. The particle is released from rest at a distance a below its equilibrium position. The motion

of P takes place in a medium which offers a resistance of magnitude

$$kmv\sqrt{\left(\frac{g}{a}\right)}$$

where v is the speed of the particle.

At time t the displacement of P below its equilibrium position is x. Show that, so long as the string remains taut,

$$a\frac{\mathrm{d}^2x}{\mathrm{d}t^2} + k\sqrt{(ag)}\frac{\mathrm{d}x}{\mathrm{d}t} + k^2gx = 0$$

Hence show that, while the string remains stretched,

$$x = ae^{-kwt}\left[\cos\left(kwt\sqrt{3}\right) + \frac{1}{\sqrt{3}}\sin\left(kwt\sqrt{3}\right)\right]$$

where $w = \sqrt{\left(\frac{g}{4a}\right)}$.

Deduce that P comes to rest before the string becomes slack provided that

$$k^2 < e^{\frac{\pi}{\sqrt{3}}}$$

16 A particle of mass m moves in a straight horizontal line and at time t its displacement from a fixed point O on the line is x. The forces acting on the particle are

(i)   a force $-2mk^2x\mathbf{i}$ where $\mathbf{i}$ is a unit vector along Ox,

(ii) a resistance to the motion of magnitude $3mkv$ where v is the speed of the particle,

(iii) a force $3mak^2\cos(kt)\mathbf{i}$,

where a and k are positive constants.

Write down a differential equation satisfied by x.

Given that at $t = 0$ the particle is at rest at O, find x at time t.

[E]

17 A particle P is attached to one end of a spring. The other end of the spring oscillates. The displacement x of P, from a fixed reference point at time t, satisfies the differential equation

$$\frac{\mathrm{d}^2x}{\mathrm{d}t^2} + 4x = \lambda\sin 4t$$

where λ is a constant. Given that $x = 0$ and $\frac{\mathrm{d}x}{\mathrm{d}t} = \frac{\lambda}{3}$ when $t = 0$ obtain x as a function of t.

18 The differential equation for the motion of a particle, which is constrained to move along the x-axis, is

$$\frac{d^2x}{dt^2} + 2k\frac{dx}{dt} + \omega_0^2 x = f(t), \ \omega_0 > k$$

where ω_0 and k are positive constants. Describe a physical problem for which this could be the differential equation. Obtain the general solution of the equation in the particular case when $f(t) = a\cos \omega t$, where a and ω are positive constants. Find the amplitude of the forced oscillation.

19 A particle P, of mass m kilograms, is fastened to one end of a light spring of natural length L metres and modulus λmg newtons, where λ is a constant. The other end is fixed to the roof of a stationary lift and P is hanging in equilibrium. At time $t = 0$ the lift starts to move vertically upwards with constant speed u metres per second and the distance of P above its initial position after t seconds is y metres.

(a) By considering the total extension in the spring after t seconds show that

$$\frac{d^2y}{dt^2} + k^2y = k^2ut$$

where $k^2 = \dfrac{\lambda g}{L}$ and $y = 0$, $\dfrac{dy}{dt} = 0$ at $t = 0$.

(b) Find y in terms of t and k.

After T seconds, where $T > 0$, a stationary observer outside the lift notices that the particle is instantaneously at rest.

(c) Find the smallest value of T.   [E]

20 A particle P, of mass m, is attached to one end of a light elastic string, of natural length L and modulus $8mn^2L$, where n is a constant. The other end of the string is attached to a fixed point O on the horizontal table on which P moves. Initially P is at rest on the table with $OP = L$. A force is now applied to P in the direction OP. The magnitude of the force is mn^2Le^{-nt} where t is the time measured from the initial application of the force. The motion of P is opposed by a resistance of magnitude $6mn$ times the speed of P. Show that the extension x of the string satisfies the differential equation

$$\frac{d^2x}{dt^2} + 6n\frac{dx}{dt} + 8n^2x = n^2Le^{-nt}$$

Find x in terms of t.

21 A uniform rod, of mass M and length l, can rotate freely in a vertical plane about a smooth hinge at one end. A light inextensible string attached to the other end of the rod passes over a smooth peg which is at a height $h(>l)$ vertically above the hinge and supports a mass m hanging freely.

Show that the potential energy of the system is given by

$$\tfrac{1}{2}Mgl\cos\theta + mg\sqrt{(l^2 + h^2 - 2hl\cos\theta)} + \text{constant}$$

where θ is the inclination of the rod to the upward vertical. Prove that the positions of equilibrium in which the rod is vertical are both stable if

$$1 + \frac{l}{h} > \frac{2m}{M} > 1 - \frac{l}{h} \qquad\qquad \text{[E]}$$

22 A light rod AB, of length $2a$, can turn freely in a vertical plane about a smooth fixed hinge at A. A particle of mass m is attached at B. One end of a light elastic string, of natural length a and modulus $\dfrac{mg}{\sqrt{3}}$, is attached to B. The other end of the string is attached to a fixed point O at the same horizontal level as A. Given that $OA = 2a$ and that the angle between AB and the downward vertical at A is $\left(\dfrac{\pi}{2} - 2\theta\right)$, show that, provided the string remains taut, the potential energy of the system is

$$\frac{-2mga}{\sqrt{3}}(\sqrt{3}\sin 2\theta + 2\cos 2\theta + 2\sin\theta) + \text{constant}.$$

Verify that there is a position of equilibrium in which $\theta = \dfrac{\pi}{6}$ and determine whether this position is stable or unstable.

[E]

23 A uniform rod AB, of mass m and length $2a$, is free to rotate in a vertical plane about a smooth hinge at A. A light elastic string, of natural length a and modulus $\dfrac{mg}{2}$, is attached to the end B and to a small ring which can move freely on a smooth horizontal wire at a height $3a$ above A and in a vertical plane through A. Show that the system cannot be in equilibrium unless the string is vertical.

If AB makes an angle θ with the downward vertical and the string is vertical, show that the potential energy of the system is V, where

$$V = mga(\cos\theta + \cos^2\theta) + \text{constant}.$$

Determine the positions of equilibrium of the system and examine their stability.   [E]

24

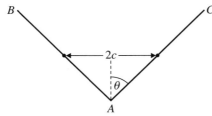

The diagram shows two uniform rods AB, AC, each of mass m and length $2a$, smoothly jointed at A and placed symmetrically, in a vertical plane, over two smooth pegs which are fixed at a distance $2c$ apart, $c < a$, on the same horizontal line. Given that each rod makes an angle θ with the upward vertical through A, show that the potential energy of the system is given by

$$2mg(a\cos\theta - c\cot\theta) + \text{constant}.$$

Hence, or otherwise, show that the equilibrium configurations occur when

$$\sin^2\theta = \frac{c}{a}.$$

Discuss the stability of the two possible positions of equilibrium.   [E]

25 A uniform rod AB, of mass m and length $2a$, can turn about a fixed smooth pivot at A. A light inextensible string is attached to the other end B of the rod, and passes through a smooth ring fixed at a point C vertically above A so that $AC = 2a$. A particle P of mass $\dfrac{m}{2}$ hangs from the other end of the string. The angle CAB is θ, where $0 \leqslant \theta \leqslant \pi$. Show that the potential energy, V, of the system is given by:

$$V = 2mga\sin\left(\frac{\theta}{2}\right) + mga\cos\theta + \text{constant}.$$

Show also that the system is in equilibrium when $\theta = \dfrac{\pi}{3}$ and when $\theta = \pi$ and investigate the stability of these positions of equilibrium.   [E]

26 A small ring of mass M can move freely on a smooth circular hoop, of radius a, fixed in a vertical plane. Two light inextensible strings attached to the ring pass over two small smooth pegs situated on the hoop at the same horizontal level below the centre of the hoop. To their free ends are attached particles each of mass m. The distance between the pegs is $2a \sin \alpha$. Given that the ring is below the pegs and that $M < m \tan\left(\frac{1}{2}\alpha\right)$, prove that the number of positions of equilibrium is three or one depending whether $M > m \sin\left(\frac{1}{2}\alpha\right)$ or $M \leqslant m \sin\left(\frac{1}{2}\alpha\right)$.

Discuss the stability of the positions of equilibrium when $M > m \sin\left(\frac{1}{2}\alpha\right)$. [E]

27 Seats on a coach rest on stabilisers to enable the seats to return to their initial positions smoothly after the coach hits a bump in the road. In a mathematical model of the situation, the following assumptions are made: each stabiliser is a light elastic spring, enclosed in a viscous liquid and fixed in a vertical position; the spring exerts a force of 1.8 N for each cm by which it is extended or compressed; the seat, together with the person sitting on it, constitute a particle P attached to the upper end of the spring which is vertical, the lower end of the spring being fixed; the viscous liquid exerts a resistance to the motion of P of magnitude $240v$ N when the speed of P is $v\,\mathrm{m\,s^{-1}}$. Given that the mass of P is $m\,\mathrm{kg}$, and the distance of P from its equilibrium position at time t seconds is x metres measured in a downwards direction,

(a) show that x satisfies the differential equation

$$m\frac{\mathrm{d}^2 x}{\mathrm{d}t^2} + 240\frac{\mathrm{d}x}{\mathrm{d}t} + 180x = 0.$$

(b) Show that, when P is disturbed from its equilibrium position, the resulting motion is oscillatory when $m > 80$.

A man is sitting on the seat when the coach hits a bump in the road, giving the seat an initial upward speed of $U\,\mathrm{m\,s^{-1}}$. The combined mass of the man and the seat is 80 kg.

(c) Find an expression for x in terms of t.

(d) Find the greatest displacement of the man from his equilibrium position in the subsequent motion. [E]

28 A particle P of mass 0.2 kg is attached to one end of a light
elastic string of natural length 0.6 m and modulus of elasticity
0.96 N. The other end of the string is fixed to a point which is
0.6 m above the surface of a liquid. The particle is held on the
surface of the liquid, with the string vertical, and then released
from rest. The liquid exerts a constant upward force on P of
magnitude 1.48 N, and also a resistive force of magnitude
1.2v N, when the speed of P is v m s^{-1}. At time t seconds, the
distance travelled down by P is x metres.

(a) Show that, during the time when P is moving downwards,

$$\frac{\mathrm{d}^2 x}{\mathrm{d}t^2} + 6\frac{\mathrm{d}x}{\mathrm{d}t} + 8x = 2.4.$$

(b) Find x in terms of t.

(c) Show that the particle continues to move down through
the liquid throughout the motion.   [E]

29 Two uniform rods AB and AC, each of length $2l$ and mass m,
are freely hinged at A and rest symmetrically over a fixed
smooth cylinder of radius a. The axis of the cylinder is
horizontal and perpendicular to the vertical plane containing
the rods. Given that the angle between the rods is 2θ and that
$2l > a\cot\theta$, show that the potential energy V of the rods is

$$V = 2mg(a\,\mathrm{cosec}\,\theta - l\cos\theta) + \text{constant}.$$

(a) Show that, for small displacements in which A moves in a
vertical straight line passing through the axis of the cylinder, a
position of equilibrium occurs when $\theta = \beta$ where

$$a\cos\beta = l\sin^3\beta.$$

(b) Show, by considering the graphs of $y = a\cos\beta$ and
$y = l\sin^3\beta$, that for $0 < \beta < \dfrac{\pi}{2}$, the equation

$$a\cos\beta = l\sin^3\beta$$

has a unique solution in this range.

(c) Determine the nature of this position of equilibrium.   [E]

Examination style paper

M4

Answer all questions   **Time allowed 90 minutes**

Whenever a numerical value of g is required, take $g = 9.8\,\mathrm{m\,s^{-2}}$

1. Bill is riding his horse at $5\,\mathrm{km\,h^{-1}}$ due North. He observes Amy riding her horse. She appears to be moving at $4\,\mathrm{km\,h^{-1}}$ in the direction $\mathrm{N}\,60°\,\mathrm{E}$. Find Amy's true velocity. **(6 marks)**

2. A particle P is released from rest at a height a above a horizontal plane. The resistance to the motion of the particle is kv^2 per unit mass, where v is the speed of P and k is a positive constant. The particle strikes the plane with speed V. Show that
$$kV^2 = g(1 - e^{-2ka})$$ **(7 marks)**

3. A light spring AB of natural length l and modulus of elasticity mln^2 lies straight and at its natural length at rest on a horizontal table. A particle of mass m is attached to the end A. At $t = 0$ the end B is moved in a straight line in the direction AB with constant acceleration f, so that at time t its displacement in this direction from its initial position is $\frac{1}{2}ft^2$.
The displacement of the particle, at time t, in the direction BA from its initial position is x.

   (a) Show that x satisfies the differential equation
$$\frac{d^2x}{dt^2} + n^2x = \tfrac{1}{2}n^2ft^2.$$ **(4 marks)**

   By using the substitution $y = \frac{1}{2}ft^2 - x$, or otherwise,
   (b) find an expression for x at time t. **(7 marks)**

4.

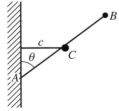

The diagram shows a uniform rod AB, of length $2a$ and a mass m, resting with end A in contact with a smooth vertical wall. The rod is supported by a smooth horizontal rail C parallel to the wall and at a distance c from the wall. A particle of mass $\frac{1}{2}m$ is attached to the rod at B.

(a) Show that, when AB makes an angle θ with the vertical, the potential energy is

$$mg\left(2a\cos\theta - \tfrac{3}{2}c\cot\theta\right) + \text{constant} \qquad \textbf{(4 marks)}$$

(b) Show that if $3c > 4a$ there is no equilibrium position, but if $3c < 4a$ there is one equilibrium position. **(5 marks)**
Given that $3c < 4a$,
(c) determine whether the position of equilibrium is stable or unstable. **(3 marks)**

5. A particle P, performing damped harmonic motion, moves along a straight line so that its displacement x metres from O at time t seconds is given by

$$x = 3e^{-12t}\cos 5t, \; t \geqslant 0.$$

The particle P, comes to instantaneous rest first at the point A. P is next instantaneously at rest at the point B. Show that

$$\frac{AO}{BO} = e^{\frac{12\pi}{5}}. \qquad \textbf{(12 marks)}$$

6.

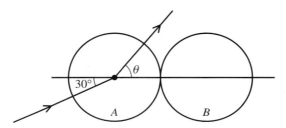

A small uniform sphere A of mass $3m$ moving with speed u on a smooth horizontal table collides with a stationary small uniform sphere B, of the same size as A and of mass $2m$. The direction of motion of A makes an angle of $30°$ with the line of centres of A and B. The direction of motion of A after the impact makes an angle θ with the line of centres as shown in the figure. The coefficient of restitution between the spheres is $\tfrac{2}{3}$.
(a) Find θ. **(9 marks)**
(b) Find the impulse exerted by A on B. **(4 marks)**

7. (In this question the horizontal unit vectors $\mathbf{i}$ and $\mathbf{j}$ are directed due East and due North respectively.)
The airport B is due north of airport A. On a particular day the velocity of the wind is $(80\mathbf{i} + 25\mathbf{j})\,\mathrm{km\,h^{-1}}$. Relative to the air an aircraft flies with constant speed $208\,\mathrm{km\,h^{-1}}$. When the aircraft flies directly from A to B
(a) show that its speed relative to the ground is $217\,\mathrm{km\,h^{-1}}$. **(7 marks)**
After flying from A to B the aircraft returns directly to A. If the time taken on the outward journey is T_1 hours and the time taken on the return flight is T_2 hours,
(b) find $\dfrac{T_1}{T_2}$. **(7 marks)**

Answers

Exercise 1A

1. (a) $(-4\mathbf{i} - 9\mathbf{j})\,\text{m}$
   (b) $(3\mathbf{i} - 7\mathbf{j})\,\text{m}$
   (c) $(-5\mathbf{i} - \mathbf{k})\,\text{m}$
2. (a) $(5\mathbf{i} + 12\mathbf{j})\,\text{m s}^{-1}$ $13\,\text{m s}^{-1}$
   (b) $(4\mathbf{i} + 3\mathbf{j})\,\text{m s}^{-1}$ $5\,\text{m s}^{-1}$
   (c) $(-3\mathbf{i} - 7\mathbf{j} + 9\mathbf{k})\,\text{m s}^{-1}$ $\sqrt{139}\,\text{m s}^{-1}$
3. $(200\mathbf{i} + 470\mathbf{j})\,\text{m s}^{-1}$
4. $1600\,\text{h}$ $(9\mathbf{i} + 4\mathbf{j})\,\text{km}$
5. (a) $60°$ to bank (upstream)
   (b) $63.5\,\text{s}$
6. $4.51\,\text{m s}^{-1}$ $85.4°$ to bank (downstream)
7. $052°$
8. $117\,\text{km h}^{-1}$ N $26.3°$ E
9. $16\sqrt{2}\,\text{km h}^{-1}$ from south west
10. $27.5\,\text{km h}^{-1}$ from $136°$

Exercise 1B

1. $103\,\text{km}$ 12.20
2. (a) $1.45\,\text{s}$ $3.48\,\text{m}$
   (b) $(-2\mathbf{i} - 0.5\mathbf{j})\,\text{m s}^{-1}$
3. (a) $2\,\text{s}$
   (b) $p = 6.5$ $q = -5.5$
   (c) $(7\mathbf{i} + 8\mathbf{j} - 12\mathbf{k})\,\text{m}$
4. (a) $32.0\,\text{m s}^{-1}$ (b) $93.7\,\text{m}$
5. (a) $034°$ (b) $2.25\,\text{km}$ (c) 2.23 p.m.
6. (a) $9.03\,\text{km h}^{-1}$ $061°$ (b) $1.43\,\text{km}$
7. (a) $40\,\text{km h}^{-1}$ (b) 10.09
8. (a) $081°$
   (b) yes (shortest distance $= 9.33\,\text{km}$)
9. (a) $217°$ (b) $6\,\text{km}$ (c) $53\,\text{min}$

Exercise 2A

1. $17.8\,\text{m s}^{-1}$, $13°$ to wall
2. $6.32\,\text{m s}^{-1}$, $14.1°$ to wall
3. (a) $(3\mathbf{i} + 2\mathbf{j})\,\text{m s}^{-1}$ (b) $27\,\text{J}$
   (c) $18\mathbf{i}\,\text{N s}$
4. (a) $(2\mathbf{i} + 3\mathbf{j})\,\text{m s}^{-1}$ (b) $18\mathbf{j}\,\text{N s}$
10. (a) 0.484 (b) $11.1\,\text{m s}^{-1}$
11. (a) $51°$
12. $15°$

Exercise 2B

1. speed of $A = \sqrt{13}\,\text{m s}^{-1}$, $\arctan(2\sqrt{3})$ to line of centres
   speed of $B = 2\,\text{m s}^{-1}$ along line of centres
   loss of K.E. $= 1\,\text{J}$
3. sphere centre A: speed $\dfrac{5V}{4}$, making angle $127°$ with AB
   sphere centre B: speed $\dfrac{V}{4}$, along AB
4. (a) $\mathbf{v}_A = (2\mathbf{i} + 4\mathbf{j})\,\text{m s}^{-1}$, $\mathbf{v}_B = (5\mathbf{i} + 2\mathbf{j})\,\text{m s}^{-1}$
   (b) $36.9°$ (c) $81\,\text{J}$
5. speed $= \dfrac{u}{2}(1 + e)^{\frac{1}{2}}$, loss of K.E. $= \frac{1}{4}mu^2(1 - e)$
6. speed of $A = \frac{1}{2}u(4\sin^2\alpha + \frac{1}{9}\cos^2\alpha)^{\frac{1}{2}}$
   speed of $B = \frac{5}{6}u\cos\alpha$
8. (b) $v_1 = a$, $v_2 = \dfrac{a}{2}$
11. $\frac{2}{5}$
12. (a) $\dfrac{u(1 + e)}{2\sqrt{2}}$
    (b) $\dfrac{u(1 - e)}{2\sqrt{2}}$, $\dfrac{u}{\sqrt{2}}$
    (c) $84°$

13 $\dfrac{2t}{1 + 3t^2}$, maximum deflection $30°$

14 $\dfrac{160}{21}$ J

Review Exercise 1

1 (c) $t = 6.7$ min $\quad$ (d) $60.9°$

2 $\mathbf{r} = (2a - a \sin \omega t)\mathbf{i} + 2a \cos \omega t\mathbf{j}$
$\qquad + (a \sin \omega t - 3a)\mathbf{k}$
$\quad \mathbf{v} = -a\omega \cos \omega t\mathbf{i} - 2a\omega \sin \omega t\mathbf{j}$
$\qquad + a\omega \cos \omega t\mathbf{k}$
$\quad t = \dfrac{2n + 1}{\omega} \dfrac{\pi}{2} \quad \min = a\sqrt{5} \quad \max = 5a$

3 $4.82\,\mathrm{m\,s^{-1}}$, bearing $275°$, $240°$

4 $\mathbf{v}_B = 5\mathbf{i} + 12\mathbf{j}$, $\mathbf{v}_C = -4\mathbf{i}$
$\quad \mathbf{r}_B = 5t\mathbf{i} + 12t\mathbf{j}$, $\mathbf{r}_C = (80 - 4t)\mathbf{i} + 240\mathbf{j}$
$\quad 16\,\mathrm{s}$, $80\,\mathrm{m}$

5 (a) $265\,\mathrm{km\,h^{-1}}$
$\quad$ (b) $343.7°$
$\quad$ (c) $43{:}53$

6 (a) $t = \frac{14}{15}\,\mathrm{s}$ $\qquad$ (b) $u = -2$

7 (a) $\dfrac{\omega^2}{2} ma^2 (9 \sin^2 \omega t + 16 \cos^2 \omega t)$
$\quad$ (b) $ma\omega^2 (9 \cos^2 \omega t + 16 \sin^2 \omega t)^{\frac{1}{2}}$
$\quad \mathbf{r} = 3a \cos \omega t\mathbf{i} + a \sin \omega t\mathbf{j} - 4a \cos \omega t\mathbf{k}$
$\quad \mathbf{r} \cdot \mathbf{r} = a^2(25 \cos^2 \omega t + \sin^2 \omega t)$

8 (a) $5\mathbf{i} - 10\mathbf{j}$ $\qquad$ (c) 18

9 (a) $\mathbf{i} - 3\mathbf{j}$

10 (c) 1.5 $\qquad$ (d) $1\frac{1}{3}\,\mathrm{s}$

11 (a) $10\,\mathrm{km\,h^{-1}}$ $\quad$ $217°$
$\quad$ (b) 12.24
$\quad$ (c) $3\,\mathrm{km}$

12 $7\,\mathrm{m\,s^{-1}}$

13 25 knots from $037°$, $143°$, 12 knots

14 $\mathbf{r}_A = 10t\mathbf{i} + 0.6\mathbf{j}$
$\quad \mathbf{r}_B = 6.4t\mathbf{i} + 4.8t\mathbf{j}$
$\quad$ (a) $[3.6t\mathbf{i} + (0.6 - 4.8t)\mathbf{j}]\,\mathrm{km}$
$\quad$ (b) $(3.6\mathbf{i} - 4.8\mathbf{j})\,\mathrm{km\,h^{-1}}$

15 $141.8°$, 221, $3930\,\mathrm{m}$, $442\,\mathrm{s}$

16 $018.6°$

19 $e = \frac{3}{5}$, K.E. lost $= \frac{1}{5} mu^2$

21 (b) $\dfrac{V\sqrt{3}}{4}(1 - e)$ $\qquad$ (c) $\dfrac{3 - \sqrt{3}}{3} = 0.42$

22 (a) $u(\mathbf{i} + 2\mathbf{j})$, $u(2\frac{1}{2}\mathbf{i} + \mathbf{j})$ $\qquad$ (b) $\frac{4}{5}$
$\quad$ (c) $20\frac{1}{4}mu^2$, $-mu(5\mathbf{i} + 2\mathbf{j})$

25 $\dfrac{\sqrt{2}}{2} \leqslant \tan \theta \leqslant \sqrt{2}$

26 (b) $\dfrac{u\sqrt{3}}{4}(1 - e)$, $\dfrac{u}{2}$ $\quad$ (c) $73.9°$

27 (a) $\dfrac{25V}{8\sqrt{13}}$ at angle $\arctan\left(\frac{24}{7}\right)$ to line
$\quad$ of centres.

Exercise 3A

1 (a) $t = \ln 7 = 1.95$
$\quad$ (b) $v = 14\,\mathrm{e}^{-1.5} - 2 = 1.12$

2 $x = \frac{4}{3} - \frac{4}{9} \ln 4 = 0.717$

3 (a) $\frac{1}{2} \ln(51)$ metres $= 1.97\,\mathrm{m}$
$\quad$ (b) $(102\mathrm{e}^{-2} - 2)^{\frac{1}{2}}\,\mathrm{m\,s^{-1}} = 3.44\,\mathrm{m\,s^{-1}}$

4 (a) $\frac{1}{3} \arctan(2)$ seconds $= 0.369\,\mathrm{s}$
$\quad$ (b) $\frac{1}{3} \ln 5$ metres $= 0.536\,\mathrm{m}$

6 $\dfrac{1}{19.6k} \ln\left(1 + \dfrac{ku^2}{\mu}\right)\,\mathrm{m}$
$\quad \dfrac{1}{9.8\sqrt{(k\mu)}} \arctan\left[u\sqrt{\left(\dfrac{k}{\mu}\right)}\right]\,\mathrm{s}$

7 (a) $\left[\dfrac{g}{k}(1 - \mathrm{e}^{-2kd}) + u^2\mathrm{e}^{-2kd}\right]^{\frac{1}{2}}$
$\quad$ (b) $(u^2 + 2gd)^{\frac{1}{2}}$

10 $6.38\,\mathrm{m}$

11 $\dfrac{m}{3k} \ln\left(\frac{63}{37}\right)$

Exercise 4A

1 $x = \mathrm{e}^{-\frac{\pi}{2}}$; damped oscillatory motion

2 $x = 4\mathrm{e}^{-t}(1 + t)$; $8\mathrm{e}^{-2}\,\mathrm{m\,s^{-1}}$

3 $x = \mathrm{e}^{-t}(2 \cos 2t + \sin 2t)$

4 $x = 2\mathrm{e}^{-t} - \mathrm{e}^{-2t}$; $t = \ln(10 + 3\sqrt{10})$

6 (a) $x = \dfrac{u}{3k}\,\mathrm{e}^{-kt} \sin 3kt$

7 (b) $x = \dfrac{V}{3n}\,\mathrm{e}^{-nt} \sin 3nt$

8 $x = \dfrac{\lambda}{6} \sin 2t + \dfrac{\lambda}{3} \sin t$

9 $x = \dfrac{V}{6n}(8 - 9\mathrm{e}^{-nt} + \mathrm{e}^{-3nt})$

10 (a) $\ddot{x} + k\dot{x} + \omega^2 x = ku$

11 (b) $\dfrac{2\pi}{5}$ s

12 (b) $x = \dfrac{l}{4}\,e^{-nt}\left(\cos\sqrt{15}nt + \tfrac{1}{\sqrt{15}}\sin\sqrt{15}nt\right)$

Exercise 5A

1 $V = -mga\sin\theta + \dfrac{2\sqrt{3}}{3}mga\sin\left(\dfrac{\theta}{2}\right) + \text{const}$; stable

3 If θ = angle between rod and vertical, equilibrium positions are given by $\theta = 0$ (unstable), $\theta = \pi$ (unstable), $\theta = \tfrac{1}{3}\pi$ (stable)

4 Unstable

5 $\theta = 0$ (unstable), $\cos\theta = \tfrac{3}{4}$ (stable)

Review Exercise 2

1 $\dfrac{g}{k^2}\,(1 - \ln 2)$

2 $\dfrac{1}{k}\ln\left(1 + \dfrac{ku}{g}\right)$; $\dfrac{g}{k}\,(1 - e^{-kT})$

4 $\dfrac{c}{k}\,\left(\ln 2 - \tfrac{1}{2}\right)$

7 $\dfrac{U}{g}\left[\arctan\left(\dfrac{V}{U}\right) - \dfrac{\pi}{4}\right]$

8 $\lambda = 0.24$

13 (a) $x = \tfrac{1}{2}a(1 + \sqrt{2})\,e^{(1-\sqrt{2})nt} + \tfrac{1}{2}a(1 - \sqrt{2})\,e^{-(1+\sqrt{2})nt}$

(b) $x = a\,e^{-\frac{nt\sqrt{3}}{2}}\left(\cos\tfrac{1}{2}nt + \sqrt{3}\sin\tfrac{1}{2}nt\right)$

14 $x = \dfrac{V}{n}\,e^{-nt}\cos\left(nt + \dfrac{\pi}{2}\right)$

16 $\ddot{x} + 3k\dot{x} + 2k^2 x = 3ak^2\cos kt$

$x = \dfrac{6a}{5}\,e^{-2kt} - \dfrac{3a}{2}\,e^{-kt} + \dfrac{3a}{10}\cos kt + \dfrac{9a}{10}\sin kt$

17 $x = \dfrac{\lambda}{3}\sin 2t - \dfrac{\lambda}{12}\sin 4t$

18 $\dfrac{a}{\left[(\omega^2 - \omega_0^2)^2 + 4k^2\omega^2\right]^{\frac{1}{2}}}$

19 (b) $y = u\left(t - \dfrac{1}{k}\sin kt\right)$

(c) $T = \dfrac{2\pi}{k}$

20 $x = L\left(-\tfrac{1}{2}\,e^{-2nt} + \tfrac{1}{6}\,e^{-4nt} + \tfrac{1}{3}\,e^{-nt}\right)$

22 stable

23 $\theta = 0$ unstable
$\theta = 120°$ stable
$\theta = 180°$ unstable

24 θ acute, unstable $\quad\theta$ obtuse, stable

25 $\theta = \dfrac{\pi}{3}$, unstable $\quad\theta = \pi$, stable

26 $\theta = 0$ stable; others unstable

27 (c) $x = -ut\,e^{-\frac{3}{2}t}$

(d) $\dfrac{2u}{3e}$

28 (b) $x = 0.3\,(e^{-4t} - 2\,e^{-2t} + 1)$

29 (c) stable

Examination style paper M4

1 $7.81\ \mathrm{km\,h^{-1}}$ at $\mathrm{N}\,26.3°\,\mathrm{E}$

3 (b) $x = \tfrac{1}{2}ft^2 - \dfrac{f}{n^2} + \dfrac{f}{n^2}\cos nt$

4 (c) unstable

6 (a) $\theta = 60°$

(b) $\sqrt{3}mu$

7 $\dfrac{T_1}{T_2} = \dfrac{167}{217}$

List of symbols and notation

The following notation will be used in all Edexcel examinations.

$\in$	is an element of
$\notin$	is not an element of
$\{x_1, x_2, \ldots\}$	the set with elements $x_1, x_2, \ldots$
$\{x : \ldots\}$	the set of all x such that $\ldots$
$n(A)$	the number of elements in set A
$\varnothing$	the empty set
$\mathscr{E}$	the universal set
A'	the complement of the set A
$\mathbb{N}$	the set of natural numbers, $\{1, 2, 3, \ldots\}$
$\mathbb{Z}$	the set of integers, $\{0, \pm 1, \pm 2, \pm 3, \ldots\}$
$\mathbb{Z}^+$	the set of positive integers, $\{1, 2, 3, \ldots\}$
$\mathbb{Z}_n$	the set of integers modulo n, $\{0, 1, 2, \ldots, n-1\}$
$\mathbb{Q}$	the set of rational numbers $\left\{\dfrac{p}{q} : p \in \mathbb{Z}, q \in \mathbb{Z}^+\right\}$
$\mathbb{Q}^+$	the set of positive rational numbers, $\{x \in \mathbb{Q} : x > 0\}$
$\mathbb{Q}_0^+$	the set of positive rational numbers and zero, $\{x \in \mathbb{Q} : x \geqslant 0\}$
$\mathbb{R}$	the set of real numbers
$\mathbb{R}^+$	the set of positive real numbers, $\{x \in \mathbb{R} : x > 0\}$
$\mathbb{R}_0^+$	the set of positive real numbers and zero, $\{x \in \mathbb{R} : x \geqslant 0\}$
$\mathbb{C}$	the set of complex numbers
(x, y)	the ordered pair x, y
$A \times B$	the cartesian product of sets A and B, $A \times B = \{(a, b) : a \in A, b \in B\}$
$\subseteq$	is a subset of
$\subset$	is a proper subset of
$\cup$	union
$\cap$	intersection
$[a, b]$	the closed interval, $\{x \in \mathbb{R} : a \leqslant x \leqslant b\}$
$[a, b)$	the interval $\{x \in \mathbb{R} : a \leqslant x < b\}$
$(a, b]$	the interval $\{x \in \mathbb{R} : a < x \leqslant b\}$
(a, b)	the open interval $\{x \in \mathbb{R} : a < x < b\}$
$y \, R \, x$	y is related to x by the relation R
$y \sim x$	y is equivalent to x, in the context of some equivalence relation
$=$	is equal to
$\neq$	is not equal to
$\equiv$	is identical to *or* is congruent to

$\approx$	is approximately equal to
$\cong$	is isomorphic to
$\propto$	is proportional to
$<$	is less than
$\leqslant$, $\ngtr$	is less than or equal to, is not greater than
$>$	is greater than
$\geqslant$, $\nless$	is greater than or equal to, is not less than
∞	infinity
$p \wedge q$	p and q
$p \vee q$	p or q (or both)
$\sim p$	not p
$p \Rightarrow q$	p implies q (if p then q)
$p \Leftarrow q$	p is implied by q (if q then p)
$p \Leftrightarrow q$	p implies and is implied by q (p is equivalent to q)
$\exists$	there exists
$\forall$	for all
$a + b$	a plus b
$a - b$	a minus b
$a \times b$, ab, $a.b$	a multiplied by b
$a \div b$, $\dfrac{a}{b}$, a/b	a divided by b
$\sum_{i=1}^{n} a_i$	$a_1 + a_2 + \ldots + a_n$
$\prod_{i=1}^{n} a_i$	$a_1 \times a_2 \times \ldots \times a_n$
$\sqrt{a}$	the positive square root of a
$|a|$	the modulus of a
$n!$	n factorial
$\begin{pmatrix} n \\ r \end{pmatrix}$	the binomial coefficient $\dfrac{n!}{r!(n-r)!}$ for $n \in \mathbb{Z}^{+}$ $\dfrac{n(n-1)\ldots(n-r+1)}{r!}$ for $n \in \mathbb{Q}$
$\mathrm{f}(x)$	the value of the function f at x
$\mathrm{f} : A \rightarrow B$	f is a function under which each element of set A has an image in set B
$\mathrm{f} : x \mapsto y$	the function f maps the element x to the element y
f^{-1}	the inverse function of the function f
$\mathrm{g} \circ \mathrm{f}$, gf	the composite function of f and g which is defined by $(\mathrm{g} \circ \mathrm{f})(x)$ or $\mathrm{gf}(x) = \mathrm{g}(\mathrm{f}(x))$
$\lim_{x \to a} \mathrm{f}(x)$	the limit of f(x) as x tends to a
Δx, δx	an increment of x
$\dfrac{\mathrm{d}y}{\mathrm{d}x}$	the derivative of y with respect to x

$\dfrac{d^n y}{dx^n}$	the nth derivative of y with respect to x
$f(x), f(x), \ldots f^{(n)}(x)$	the first, second, $\ldots$ nth derivatives of $f(x)$ with respect to x
$\displaystyle\int y\, dx$	the indefinite integral of y with respect to x
$\displaystyle\int_a^b y\, dx$	the definite integral of y with respect to x between the limits $x = a$ and $x = b$
$\dfrac{\partial V}{\partial x}$	the partial derivative of V with respect to x
$\dot{x}, \ddot{x}, \ldots$	the first, second, $\ldots$ derivatives of x with respect to t
e	base of natural logarithms
e^x, $\exp x$	exponential function of x
$\log_a x$	logarithm to the base a of x
$\ln x$, $\log_e x$	natural logarithm of x
$\lg x$, $\log_{10} x$	logarithm to the base 10 of x
sin, cos, tan cosec, sec, cot	the circular functions
arcsin, arccos, arctan arccosec, arcsec, arccot	the inverse circular functions
sinh, cosh, tanh cosech, sech, coth	the hyperbolic functions
arsinh, arcosh, artanh, arcosech, arsech, arcoth	the inverse hyperbolic functions
i	square root of -1
z	a complex number, $z = x + iy$
Re z	the real part of z, Re $z = x$
Im z	the imaginary part of z, Im $z = y$
$|z|$	the modulus of z, $|z| = \sqrt{(x^2 + y^2)}$
arg z	the argument of z, arg $z = \arctan\dfrac{y}{x}$
z^*	the complex conjugate of z, $x - iy$
$\mathbf{M}$	a matrix $\mathbf{M}$
$\mathbf{M}^{-1}$	the inverse of the matrix $\mathbf{M}$
$\mathbf{M}^{T}$	the transpose of the matrix $\mathbf{M}$
det $\mathbf{M}$, $|\mathbf{M}|$	the determinant of the square matrix $\mathbf{M}$
$\mathbf{a}$	the vector $\mathbf{a}$
$\overrightarrow{AB}$	the vector represented in magnitude and direction by the directed line segment AB
$\hat{\mathbf{a}}$	a unit vector in the direction of $\mathbf{a}$
$\mathbf{i}, \mathbf{j}, \mathbf{k}$	unit vectors in the directions of the cartesian coordinate axes
$|\mathbf{a}|$, a	the magnitude of $\mathbf{a}$
$|\overrightarrow{AB}|$, AB	the magnitude of $\overrightarrow{AB}$

$\mathbf{a} \cdot \mathbf{b}$	the scalar product of $\mathbf{a}$ and $\mathbf{b}$
$\mathbf{a} \times \mathbf{b}$	the vector product of $\mathbf{a}$ and $\mathbf{b}$
A, B, C, etc	events
$A \cup B$	union of the events A and B
$A \cap B$	intersection of the events A and B
$\mathrm{P}(A)$	probability of the event A
$\overline{A}$	complement of the event A
$\mathrm{P}(A|B)$	probability of the event A conditional on the event B
X, Y, R, etc.	random variables
x, y, r, etc.	values of the random variables X, Y, R, etc
$x_1, x_2 \ldots$	observations
$f_1, f_2, \ldots$	frequencies with which the observations $x_1, x_2, \ldots$ occur
$\mathrm{p}(x)$	probability function $\mathrm{P}(X = x)$ of the discrete random variable X
$p_1, p_2, \ldots$	probabilities of the values $x_1, x_2, \ldots$ of the discrete random variable X
$\mathrm{f}(x), \mathrm{g}(x), \ldots$	the value of the probability density function of a continuous random variable X
$\mathrm{F}(x), \mathrm{G}(x), \ldots$	the value of the (cumulative) distribution function $\mathrm{P}(X \leqslant x)$ of a continuous random variable X
$\mathrm{E}(X)$	expectation of the random variable X
$\mathrm{E}[\mathrm{g}(X)]$	expectation of $\mathrm{g}(X)$
$\mathrm{Var}(X)$	variance of the random variable X
$\mathrm{G}(t)$	probability generating function for a random variable which takes the values 0, 1, 2, $\ldots$
$\mathrm{B}(n, p)$	binomial distribution with parameters n and p
$\mathrm{N}(\mu, \sigma^2)$	normal distribution with mean μ and variance σ^2
μ	population mean
σ^2	population variance
σ	population standard deviation
$\bar{x}, m$	sample mean
$s^2, \hat{\sigma}^2$	unbiased estimate of population variance from a sample, $$s^2 = \frac{1}{n-1}\sum(x_i - \bar{x})^2$$
ϕ	probability density function of the standardised normal variable with distribution $\mathrm{N}(0, 1)$
Φ	corresponding cumulative distribution function
ρ	product-moment correlation coefficient for a population
r	product-moment correlation coefficient for a sample
$\mathrm{Cov}\,(X, Y)$	covariance of X and Y

Index